"Clear, straightforward, lucid, helpful. Highly rec _____ ... Iou and your organization will be better off for it. Authors Hung Le and Grace Duffy present a practical approach for achieving operational excellence based on culture first, process and tools second. The authors outline how tools support processes that work systematically to accomplish the organization's strategic goals, but everything begins with a healthy culture and mindset."

Christopher Welker, *Vice President and Chief Information Officer, Northrop Grumman Space Systems*

"Finally, an outstanding book that introduces Lean Six Sigma in a manner that will be appealing to everyone from front-line supervisors to CEOs. Additionally, this will be an excellent textbook for college business students. The authors skillfully focus on the people side of Lean Six Sigma, introducing tools in the book's later chapters. Very well organized with easy-to-understand discussion and captivating stories, it covers fundamentals that everyone needs to know before they get immersed in the details, acronyms, and jargon of Lean Six Sigma."

Colonel Robert R. (Bob) Sarratt, *U.S. Army, Retired; President, Sarratt Acquisition Management Inc.*

"If you interact with people at all, this book is a must to help you develop your personal plan of intentional design and approach with people. I found clear and articulate examples laid out that anyone can use to apply these techniques in any venue and at any level of an organization. Grace Duffy's and Hung Le's outtakes throughout the book were enlightening and provided relevant examples of our current workplace. The bottom line is that I learned a LOT! I wrote a lot of notes and already have ideas and plans as to how I will adapt and incorporate these learnings into my current work and life processes. I hope you will get equal enjoyment and reap the same benefits!"

Jd Marhevko, *Vice President, Quality ZF Group ADAS & Electronics*

"This book provides powerful and practical advice on taking an organization to the next level. It explains how shifting behaviors of the entire team, including management, will unlock the potential of the organization while establishing a positive culture. A must-read for anyone wanting to improve performance."

Patricia Gillis, *Vice President Quality, Aerospace & Defense Industry*

Human-Centered Lean Six Sigma

This book focuses on the human side of organizational culture. The authors approach organizational culture from the perspective of alignment to mission, vision, and values. Using a Lean Six Sigma structure, the sequence of chapters begins with the organization and its structure, then drills through strategic, operational, and tactical levels of process and behavior which establish and grow the overall culture of the organization over time. The book begins with foundational principles of organization, through the necessity of aligning processes and systems to mission and vision, assessment, gap analysis for improvement, prioritization, and chapters on qualitative and quantitative approaches for reducing variation and improving systems and behavior.

Through this book, readers will:

- Learn the foundation and core concepts of the organization
- Discover the "right" focus of shifting the culture of the organization
- Recognize the building blocks of organizational culture and how to integrate them into a successful, customer-focused system of interconnected processes
- Focus on people as drivers of technology, rather than the reverse
- Explore techniques to address the challenges and concerns of today's training and deployment for organizational performance excellence
- Use the chapters as short discussions or training workshops for either internal education or public/private technical education.

Human-Centered
Lean Six Sigma
Creating a Culture of Integrated
Operational Excellence

Hung Le and Grace Duffy

A PRODUCTIVITY PRESS BOOK

First published 2024
by Routledge
605 Third Avenue, New York, NY 10158

and by Routledge
4 Park Square, Milton Park, Abingdon, Oxon, OX14 4RN

Routledge is an imprint of the Taylor & Francis Group, an informa business

ISBN: 978-1-032-59484-2 (hbk)
ISBN: 978-1-032-59483-5 (pbk)
ISBN: 978-1-003-45489-2 (ebk)

DOI: 10.4324/9781003454892

Typeset in ITC Garamond Std
by KnowledgeWorks Global Ltd.

Contents

List of Figures and Tables

Figures

Tables

Foreword

Dr. Hung Le and Grace Duffy have used their extensive knowledge and experience in people and quality management to create a captivating book on how to energize people to pursue excellence. While both worked in the corporate quality arena, their paths finally crossed in the early 2000s. Since then, they have worked together to teach Lean Six Sigma classes and made joint presentations at several ASQ World Conferences on Quality Improvement. The outstanding receptivity to their classes and presentations led them to develop this book.

Forty-eight years ago, Hung's grandmother asked me if I could help Hung, his sister, brother, and aunt to leave South Vietnam as it was being overrun by the North Vietnamese forces. I was an Army Major with the U.S. Defense Attaché office and had known Hung's family for years. I put them on one of the evacuation flights, and several months later, they arrived at my home in Virginia.

- When I took 14-year-old Hung to the high school in the summer of 1975 to enroll him in the 9th grade, they wanted to have him repeat the 8th grade until he amazed them with his knowledge of math. He graduated in three years (after the 11th grade), finished his B.S. in 3-1/2 years, and his MS, Ph.D., and MBA in rapid succession, all on his own, without any help from me.
- After starting his career at IBM, Dr. Le soon focused on Lean Six Sigma, advancing to key positions with Northrop Grumman, where, as a Six Sigma Master Black Belt, he provided numerous workshops and Kaizen Events while training over 2,500 Yellow Belts and 300 Green Belts. He helped stand up the Six Sigma Program and provided Master Black Belt support and deployed Six Sigma at two major Divisions. Over the years, Dr. Le has facilitated numerous Affordability/Lean Workshops and Kaizen Events for new captures and programs to create and deliver more value for customers.

■ Dr. Le has also served as Adjunct Professor in Operations Research and Statistics at George Mason University and has authored and/or presented over 40 technical papers, primarily in the areas of statistical signal processing, modeling and simulation, and quality and change management.

Grace focused her career on quality, management, and education.

■ Like Dr. Le, Grace used her technical, educator, management, and leadership experience at IBM to springboard into a variety of different pursuits.
■ As Business Department Head and Dean for Economic Development at Trident Technical College, Grace worked to ensure the 10,000 students were prepared to meet the needs of business in quality, economics, marketing, finance, management, and leadership. Her service as President and CEO of the Trident Area Community of Excellence allowed her to support CEOs of major corporations in the South Carolina Low Country in process benchmarking and organizational excellence.
■ For 30 years as President of Management and Performance Systems, Grace has used her anthropology degree, as well as her MBA to develop approaches for organizational design, human-centered culture, and strategic alignment to assist executives from private and nonprofit companies improve results.

Hung's and Grace's unique backgrounds in industry management, industry training programs, and academic teaching have enabled them to develop a book that builds first on the human side of Lean Six Sigma and then introduces the tools used to plan, track, and achieve success in process improvement. Readers should find this approach much more interesting than jumping feet first into the details of the Lean Six Sigma tools. Universities will find this book an excellent graduate business and systems management text, designed to support a semester class on organizational culture and design.

Looking back on my years in the Department of Defense and industry managing clothing and individual equipment research, development, test, engineering, production, and product quality, I wish that we would have had this book, *Human-Centered Lean Six Sigma: Creating a Culture of Integrated Operational Excellence*. It would have made my life a lot easier in training our workforce and increased our chances of success. I expect that it will help you too.

Robert R. (Bob) Sarratt, Colonel U.S. Army, Retired
President, Sarratt Acquisition Management Inc.
Manassas, VA, April 18, 2023

Preface

I almost died a few years ago.

On an international charity trip in Southeast Asia with my wife, I contracted bacterial meningitis. I was immediately admitted to a local hospital where the quality of care was subpar from the moment I arrived. It was 90 degrees outside, and possibly inside, as well – there was no air conditioning. Patients were warehoused, two per twin bed, stacked head-to-toe. Unlike most of the locals, I could at least afford my own bed. But this was a small victory. My condition worsened despite the hospital's attempts to treat me.

My wife told my doctor that we wanted to move me to an international hospital.

The doctor informed her, "There is no need to move your husband. We have handled numerous, similar cases before."

But my wife insisted. Her decision to transfer to an international hospital, with a much higher standard of care, kept me alive.

We learned that the local hospital where I first stayed had a terrible record. Yes, they had indeed handled "numerous, similar cases" before – many such patients had checked in, but very few were able to check out!

I'm happy to report that under the care of the international hospital, I made a full recovery, and my wife and I returned safely to the United States.

The assurances of the local hospital, that they had handled cases like mine before, illustrated an important difference between a "to-do" list of prescribed activities and the outcome of all those activities. As a patient (the customer), I did not, and should not have needed to, care much about how the service was delivered or how many times the service had been rendered. I only cared about the OUTCOME. The outcome for many patients before me was death. If I had stayed in that hospital, I would not be here today.

Low Touch, High Frequency

Previously, I filled various leadership roles at IBM, GE and Northrop Grumman and consulted for companies across several industries. But that life-changing experience created an indelible shift in how I viewed service quality and the tremendous impact it has on peoples' lives, no matter what ethnicity, gender, geography or social background they come from. This viewpoint is my reason to write a book about effective ways to improve product and service quality from the ground up in any given organization, so that OUTCOMES are improved for customers. This means infusing the organization with not just improved processes, but mindful practices, resulting in an entity that hears the voice of the customer clearly from the front-line employees to the c-suite executives.

This book consists of a compilation of twelve in-person lectures branded using Stephen Covey's famous Habit #7 "Sharpen the Saw" Series, which was first piloted at a large hospital system in the United States around 2010. I presented a one-hour lecture every month, over a period of twelve months. This relatively small amount of time spent on a regular basis produced some remarkable results.

This pilot was conducted at a pharmacy department of 57 people: one director, two managers, four clinical pharmacists, 25 pharmacists, and 25 technicians. Their educational levels ranged from a high school degree to a doctorate in pharmacy.

I started with my focus not on any business metrics improvement, but rather, purely on the engagement of employees. Specifically, I began with Concepts in Organizational Alignment and Team Performance Behaviors.

In 2010, only four months into the pilot, a Gallup survey showed marked improvement in numerous employee engagement areas. Compared to the year before (2009), the following changes were noted:

■ Engagement score went from the 40th percentile to the 89th percentile, a whopping 122% improvement.
■ Staff with Perfect Attendance score went from 13 to 31, a 138% improvement.
■ Staff Sick Calls decreased 52%.
■ Employee Giving Campaign Participation went from 40% to 89%, a 122% increase.

As I continued with my lecture series, the service metrics improved as well:

■ Heart Failure Discharge Compliance went from 58.8% to 90.3%, a 54% increase.
■ Anticoagulation Service Monitoring and Intervention went from 643 to 3746, a 482% increase.
■ Pharmacists monitoring/intervening on medication orders received went from 6767 to 7872, a 16% improvement.

Besides the above improvements, a few new services were launched successfully with no additional resources. Other improved business metrics included:

■ Pharmaceutical Expenditures were reduced by $2.7M (23%)
■ Productivity improved by 22%.

This work and its results were presented at the 2011 American Society for Quality (ASQ) Conference on Quality Improvement in Pittsburg.

Having proven that we could shift a small organization's culture, I looked for ways to scale this approach. I proposed the same training model and rolled it out at a large company. Since employees there were so geographically dispersed and working on remarkably diverse programs, the classes were designed to be online once a week over twelve weeks. The time commitment was fairly minimal, and the impact was notable compared to other training models where a significant time commitment was required (i.e., 4 hours to 5 days).

I call this "high-frequency and low-touch" in our training model, and most notably, this gave me numerous opportunities to reinforce concepts and lessons learned, as opposed to the "low-frequency and high-touch" method of other models. Figure P.1 is a word map of some comments returned by class members. Class size ranged from 150 to over 450 participants. For measures, I opted to monitor leading indicators exhibiting behavior changes. Over six years, from 2014 to 2020, I trained more than 3000 workers. The training was very well-received, and such a class was promoted from word-of-mouth.

The approach I piloted addressed many of the challenges and problems facing organizations trying to develop a culture of excellence. This broad and complex topic benefits from deep insights and experiences, thus I have

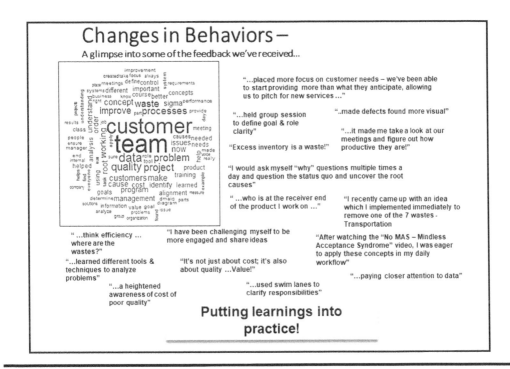

Figure P.1   The picture showing many examples of positive feedback we received from participants in the training program.

invited and collaborated with another expert in this field, Grace Duffy, with whom I have had the benefit of working, and learning from. As we share our combined passion and experience about this topic, we hope to inform and inspire our readers.

Hung Le

Acknowledgments

Many thanks to Christopher Welker, Colonel Robert R. (Bob) Sarratt, Jd Marhevko, and Patti Gillis, who took the time to read our draft manuscript. They shared excellent improvement feedback and kind words of encouragement for the value of our message.

We are grateful to Northrop Grumman for allowing us to use some of the graphics from its continuous improvement program training materials. This has allowed us to bring to life some of the best practices that are deemed crucial toward building a culture of excellence.

We wish to thank Dr. R. Scott Bonney for his input into the flow of this work. Scott spent time with us discussing the framework for a human-centered approach to the 12 Building Blocks of Organizational Culture.

Finally, we are most grateful to Michael Sinocchi at Taylor & Francis Group and Meeta Singh at KnowledgeWorks Global Ltd., for recognizing the importance of organizational culture in employee engagement. Their support in guiding us through this effort has been critical.

About the Authors

Hung Le, PhD, enjoys solving complex organizational problems with the goal of improving enterprise efficiency and effectiveness. Over the past 35 years, he has held various leadership roles at IBM, KPMG Consulting, Deloitte & Touche, GE, and Northrop Grumman, and consulted for Fortune 100 companies across a wide range of industries, including high-tech, banking, and healthcare. Before his retirement from Northrop Grumman in 2022, he ran the Program Management Office and led the Digital Transformation Initiative for the Corporate Functions and other Services Lines within the Enterprise Services Sector.

Hung Le holds a B.S. and M.S. in Electrical Engineering from the University of Virginia and the University of Maryland, respectively. A PhD in Statistical Science from George Mason University and an MBA from the Wharton Business School of the University of Pennsylvania, Dr. Le has served as Adjunct Professor in Operations Research and Statistics at George Mason University and has authored and/or presented over 40 technical papers, primarily in the areas of statistical signal processing, modeling and simulation, and quality and change management. He has also served as a board member for several non-profit organizations.

 **Grace Duffy**, President, Management and Performance Systems, has over 45 years of experience in successful business and process management in corporate, government, education, and healthcare. She provides services in organizational and process improvement, leadership, quality, customer service, and teamwork. She designs and implements effective systems for business and management success. Grace uses her experience as President, CEO, and senior manager to help organizations improve. She has authored 18 texts, additional book chapters, and many articles on quality, leadership, and organizational performance. She is a frequent speaker and trainer.

Grace holds an MBA from Georgia State University and a Bachelor's in Archaeology and Anthropology from Brigham Young University. She is an American Society for Quality (ASQ) Certified Manager of Quality/ Organizational Excellence, Certified Quality Improvement Associate, Certified Quality Auditor, and Lean Six Sigma Green Belt. Grace is an LSS Master Black Belt, ASQ Fellow, and Distinguished Service Medalist. She is the recipient of both the Quality Magazine Professional of the Year and the Asia-Pacific Quality Organization International Woman in Quality Medal. She is a member of the ISO TC 304 WG 5 Hospital Quality Management Systems and Past Chair of the ASQ Healthcare Quality and Improvement Committee, which published the U.S. Healthcare Quality Management System model and subsequent four monograph series. Grace also wrote the American National Standards Institute technical report on the U.S. Healthcare Quality Management System.

Introduction

Every day, we have opportunities to make a difference. We must seize these moments, even if we must go against the grain to do it. The problems we encounter can vary widely, from the obvious and simple to extraordinarily complex issues permeating an entire organization.

The biggest challenge in effectively addressing a problem is knowing where to start. Without understanding that, we might begin with the wrong end of the spectrum, tackling significant enterprise issues that can be unattainable or unsolvable, or homing in on small and remedial problems that have insignificant impacts on the business. We have observed many enterprise-wide initiatives misaligning with smaller initiatives, or in other words, broad intentions from the corporate penthouses utterly failing to address what's happening on the ground. Hopes for improvement are hijacked.

Therefore, a critical task is understanding how the small issues connect to the broader issues for the business. When we attain this understanding, we have a game plan: prioritizing the various challenges and executing them so that short-term gains can be achieved, sustained, and aligned to the longer-term goals. That is when we make a significant impact.

Further misalignment arises when new initiatives are launched without a clear understanding of what was done in the past, and how these new initiatives could benefit from building on past initiatives. Such initiatives complement each other when explored correctly. They can supplement and build on previous accomplishments by leveraging previous results, building on those skills and resources to speed up the launch and execution of a new initiative.

Such opportunities for improvement come with their own challenges. Organizations must be prepared and provide the time and resources, as well as the training needed, for employees to address the problems they face. Remember, however, that training alone may not suffice; we must embed mindful practices into every employee's job and at-work activities. This requires a different way of building, developing, and nurturing a culture of excellence. As Aristotle put it: "Excellence is not an act, but a habit."

When Culture Eats Strategy

Nearly 100 years have passed since Dr. Shewhart developed statistical process control in the 1920s; more than 75 since Dr. Feigenbaum first published the term "Total Quality Control." The so-called quality revolution is in its third generation. After all these decades, terms like Lean, Six Sigma, Total Quality Management, Business Process Reengineering, and the Malcolm Baldrige Performance Excellence Program have become white noise in the Board Room, only to be replaced by newer, cooler, and often more expensive and less capable versions, like digital transformation, robotic process automation, and a whole host of "intelligence" and "learning" approaches.

We make no judgment for or against these approaches, but we ask every reader one simple thing: Have you successfully implemented any of these methodologies? Your organization may have purchased training, software, and/or consultants at considerable expense, only to find that a few months down the road, change is minimal or impermanent. You see, if change isn't fostered on a cultural level from the bottom up, then implementation is either just theoretical, or it remains isolated, or both, and there it falters. Many employees never see any change and can even be unaware that anything "new" was ever tried.

We see it time and again: after years of implementing these methods, tools, and mental models, the cultures of many organizations have yet to change enough to positively impact organizational performance.

Do not just take our word for it. Talk to the old-timers in your organization. Is the quality of their work life better? Are they more engaged, appreciated, and happier coming to work each day? Do they like their coworkers more, have better friendships, and feel more respected and valued? Is employee retention up? Is it easier than ever to reach out and find qualified candidates competing for your limited job openings? Why is it such a challenge to do the things we must do every day to get results? Do we even know what "results" we should look for; what matters? Why do we say one thing and then do another?

Without effective leadership, "bad" habits form over time, negatively impacting organizational performance. As Stephen Covey famously commented, "We're but a creature of habits."[1]

There is much focus on tool usage in quality texts, with little emphasis on the behavioral aspects of learning the tools. Although leadership and quality tools are recommended and described in this book, the reader will notice our point is rarely about the tool. The tool is the means to the end, part of a holistic approach to organizational performance. A central theme of this book is that tools support processes that work systematically to accomplish the organization's strategic goals. This is not a book about tools to work on a Six Sigma project, but a recommended set of practices for everyday activities that can be sustained, ultimately generating a visible and positive shift in the organizational culture.

Leaders and improvement professionals must move from just teaching tools to instilling appropriate behaviors. We must move beyond training and goal setting to changing organizational behaviors at the grass-roots level. Culture comes from doing, not just learning. As Dr. Marc Bard commented in 2001 (and later famously quoted and popularized by Peter Drucker), "Culture eats strategy for breakfast, every day, every time.[2]"

This is all about changing and improving work habits through good practices with positive reinforcements. Intentional change takes more than brains; it takes shoe leather and sweat! Forming a Community of Practice is vital to institutionalizing changes. An old proverb states, "Sharp people sharpen one another, just as iron sharpens iron." Leadership by example and positive peer pressure can be as powerful as the tides. By focusing on the "right" habits, we can create the "right" foundational environment so positive practices reinforce and eventually become an unconscious part of one's habits.

Changing a culture will require discipline and a willingness to change at all levels of the organization. To achieve a "We" culture, we first need everyone at all levels to hold themselves accountable for their actions and how they impact the organization. Each worker must start with themselves (the "I" culture).

The twelve chapters in this book provide specifics about the failures of these implementations to make a positive culture shift.

Our Goals

The practices described in this book come from over 90 years of experience and thousands of projects we have directed and managed for clients at numerous Fortune 500 companies. Readers may notice that both authors share our stories and experiences and tell them in first-person, so don't be confused if suddenly your narrator seems to have had two different industry-spanning careers. We're a team in this endeavor!

Over the course of this book, we will promote the following goals:

Goal 1: Utilize the Best Practices of Lean Six Sigma

For those familiar with Lean Six Sigma ("LSS"), we have "unbundled" this time-tested problem-solving approach so that one does not need to solve a problem entirely to benefit from LSS tools. The components of LSS are flexible. Deconstructing a set of principles and practices may improve the effectiveness of the rollout, as many of these practices can be applied independently and need not be part of an end-to-end problem-solving process.

These unbundled concepts can be applied at any point in the decision-making process. One need not be an LSS practitioner to apply some of these best practices. Because we are faced daily with so many problems and/or decisions, the concepts brought forth in LSS allow any individual to apply them to solve any problem more effectively or make a better decision.

Goal 2: Create Sustainable Organizational Improvement

We propose a set of building blocks that form and guide, creating sustainable organizational excellence. In the past, we have witnessed waves of quality movements, such as the craftsman/apprentice system of the Middle Ages, quality inspection during the industrial revolution, quality sampling and control plans during the 20th century, leading to Total Quality Management, quality planning and now systems thinking to design excellence into a product or service at the start. Many of these methodologies consist of loosely coupled tools and techniques that can be applied to solve myriad problems.

However, overly enthusiastic organizations may dive into implementing a methodology without a deep appreciation for the state of its maturity. Lack of process control and the failure to use key performance measures renders the implementation ineffective. We've seen many LSS implementations fail to sustain any improvements because those improvements were made across different pockets of the organization when more consideration should have been given to the maturity of the overall organization.

Goal 3: Integrate Various Models to Demonstrate Application of Their Components

The chapters in this book integrate several mental constructs to help connect disparate models. This should give readers a broader understanding of applying their various tools and techniques effectively.

The chapters are written as standalone modules that can be used in an academic or team setting to assign as pre-reads. Each chapter ends with three questions to prompt the reader to organize their thoughts around the subject of the chapter. These questions make a good instructor platform for facilitating discussion and learning. By the end of the twelve chapters, the individual reader or learning group will have traveled from strategic organizational culture and alignment through assessment, data gathering, analysis, techniques, tools, measures, and finally, a return to a view of the effective organization.

Goal 4: Ensure that Lean Six Sigma Principles Permeate the Organization

LSS, by design, is human centered as it focuses on developing a deep understanding of the customers' and stakeholders' (which also includes employees') needs and values. Our objective is to ensure that LSS is driven deeply and permeated throughout the organization, not just by LSS practitioners inside (such as employees) and outside (such as consultants) the organization.

Every single employee should be an effective problem solver, supporting his/her department's goals and aligning with the organization's broader goals. As designers of this human centered LSS approach, we incorporate and leverage our own perspectives and experiences from implementing LSS so that readers may lead their improvement initiatives and attain better outcomes. We want employees to be more effective, efficient, and inspired by having access to the right tools and techniques, improving their chances of success. These employees are more productive and have higher morale. In the end, they build a strong sense of trust, and a sense of ownership in the services/products and processes fostered in their work and the people they work with.

Notes

1. Covey, Stephen R., *The 7 Habits of Highly Effective People*, Fireside Simon and Schuster, New York, 1989, pp. 47–48.
2. The Essential Drucker: The Best of Sixty Years of Peter Drucker's Essential Writings on Management (Collins Business Essentials) July 22, 2008 by Peter F. Drucker (Author).

Chapter 1

People and Culture

Bring your whole self to work. I don't believe we have a professional self Monday through Friday and a real self the rest of the time. It is all professional and it is all personal.

Sheryl Sandberg

Focusing on the Individual within Organizational Culture

"Culture is what people do when no one is watching!" That was something Jack Welch discussed at an interview in the late 90s. Maybe it's an overly simplistic view of what culture is, but when you think about it, that makes a lot of sense. Culture is what people normally do without oversight.

If we dig a little deeper into what forms an organization's culture, we find it is a combination of attitudes, beliefs, values, expectations, knowledge, language, opportunities, and structure. Culture is the cumulative result of the combination of these elements over time; it is dynamic, and it changes as the organization grows and improves itself (or, as it deteriorates, unfortunately). An organization's culture arises from the attitudes and self-confidence of each member. How members view themselves as contributing parts of the organization sustains the culture over time. Figure 1.1 suggests that individuals must be comfortable with themselves before they will be comfortable with others.

An organization's culture sets the overall climate of the organization and, moreover, becomes a major factor in employees' desire to join, stay, and grow there. It has a strong impact on employee engagement.[1] And

DOI: 10.4324/9781003454892-1

Figure 1.1 Understand and love yourself first. A stable individual is critical to an effective workplace.

employee engagement must never be underrated, as engaged employees have lower stress, turnover rates, and absenteeism, and higher commitment and productivity than those employees who are just there serving as warm bodies, waiting for a paycheck at the end of the week.

Building and sustaining a human-centered culture that encourages employee engagement is a complex endeavor requiring sound decisions. Organizational leadership must balance the strength of individual performers with the overall alignment of the organization toward the desired outcomes of the business. This engaging culture must support the strategic goals of the organization, which we will discuss further in Chapter 2.

Cultural Mindfulness and the Human-Centered Organization

Mindfulness as a concept is pervasive in today's business literature and generally, in all aspects of our conscious lives. We are encouraged from every direction to draw our attention to the moment at hand for the sake of our mental health and well-being – and small wonder, seeing how many distractions our modern world puts in our way.

Mindfulness is the purposeful non-judgmental practice of attention to the present moment. Tied to organizations, it means developing the vision, strategies, and initiatives that might guide the organization's transition to the new, sustainable business paradigm.[2] A more detailed discussion can be found in Chapter 11.

Mindful members of an organization can effectively navigate complex and fast-paced business environments by staying grounded in their values and principles, leading with compassion and empathy, and fostering a culture of growth and collaboration within the organization. Positive thinking and inner peace are critical elements of motivation. Organization leaders must be mindful of, first, how individuals contribute to the goals of the organization and, second, how the organization supports the individuals within it.

Being mindful includes exercising compassion and empathy for the individuals who perform the tasks required for customer-desired outcomes. Aligning organizational goals, external customer requirements, and internal customers (employees and stakeholders) requires not only excellent process and systems design but also human-centered design.

Design thinking is a core way of starting the journey and arriving at the right destination at the right time. We define a human-centered organization as "an organization that seeks to deliver the most value to the end customer while unlocking hidden potential and leveraging talents from every single employee to thrive in this very dynamic, fast-paced, and complex business environment."

Unless an organization is designed to be human-centered, the culture will never develop properly, and the organization will remain unable to reach its full potential.

Alignment of Culture

An effective culture aligns goals, values, and beliefs. Beliefs are internal representations of what individuals think is true, including representations of events.

Grace Duffy: *I remember how culture impacted behavior during my 20 years at a Fortune 100 technology company. Those of us who wanted to advance in the company showed incredible hustle. We talked about giving 120%, bought the latest "Dress for Success" books, enrolled in Dale Carnegie professional development courses, or joined Toastmasters to be certain we were crisp in our presentations to management. I personally made a point of enrolling in a Master of Business Administration program along with working*

60-hour weeks as a first-line supervisor. I still have those blue suits, white shirts, and Ferragamo® pumps.

The combined influence of all organizational personnel has a powerful impact on the culture as individuals learn from each other and eventually may adopt similar beliefs and attitudes. This engagement is foundational to the building of organizational culture. I was surprised when I left corporate America to serve as a Department Chair at a technical college to be corrected brittlely when I called a "staff" meeting for my faculty. I was told in no uncertain terms that "staff" was administrative, whereas "faculty" was academic. Never the twain shall meet. I learned quickly.

Building a Culture around Values

Has anyone ever asked what you value most? It's not such an easy question to answer. We may respond quickly by saying the obvious responses, like "family, success, security, and health." These answers are likely true, but they have quite different meanings for individuals.

What does "success" mean? Is it a huge bank account, a title behind your name, a difficult accomplishment, or being a recognized authority? What is "health?" Is it perfect physique, access to plentiful and nutritious food, the ability to afford medical care, a successful knee replacement surgery, or a second chance?

This is not to say that certain values are better than others but only to suggest that even when talking to ourselves, we may not always acknowledge the root of our values. Mindfulness about our real values lets us understand better what we are working for and why, *and* to adjust when necessary. For example, if we discover that we are measuring our success based on whether our peers are impressed with us, we may want to rethink either our values or our peers or both!

Organizations face the same challenge. An organization's values are often displayed in corporate lobbies for all to see so that people can mimic the behaviors that reinforce the values. It takes soul-searching to determine what those important things are that everyone in the organization must live by – and an implied agreement that, within the culture, those values carry the same meaning at each level. When leadership has a different definition of "success" or "integrity" than the mid- or front-line workers, the culture will suffer from that dissociation.

Values represent what is non-negotiable for the company. They are typically included in employee onboarding documents to ensure that new employees are aware of the expectations for behavior and deportment within the organization. Statements of values may include commitment, excellence, innovation, communication, and teamwork, among others.

Having a values statement is important but insufficient to ensure alignment between company values and managers' and employees' actions. Managers at all levels should demonstrate these values as they go about daily business.[3]

An effective organizational culture features specific characteristics. These include, but are by no means limited to:

■ Visible, engaged, and unwavering senior management support for strategically aligned initiatives.
■ Articulated vision and values.
■ Active and ongoing engagement with customers to continually identify and address current and evolving needs.
■ Clearly stated organizational goals.
■ Performance expectations for all individuals in the company which are linked to strategic and operational goals.
■ Appropriate incentives – which can favor monetary or recognition-based awards, depending on individual circumstances.

Culture is an intangible part of the organization, whether at the individual, team, department, or companywide level. The value of shifting the company unit toward a more positive culture can be substantial.

Grace Duffy: *I had the opportunity to highlight how values align with operational goals and performance expectations when teaching a Supervisory Skills class at the Ralph H. Johnson Veterans Administration Medical Center in Charleston, SC, in the early 2000s. The Medical Center supervisors and I were discussing alignment with the goals and values of the Center when the Environmental Services third shift supervisor spoke up. "How can I demonstrate to my third shift personnel that washing and waxing the hallways every night aligns with the VA and Medical Center goals and values?"*

I opened the man's question to the rest of the class, whose brainstorming culminated in a "Eureka!" moment for the Environmental Services supervisor. I summarized the conversation by saying,

Each morning when visiting hours begin, veterans' families and friends enter the center to visit their loved ones. Your third shift personnel, by providing clean floors with a dry, shiny wax finish, present a bright and positive atmosphere for these families and friends. What a pleasant greeting to give to those who visit our veterans each day.

The Environmental Services supervisor was visibly touched by this contribution his third shift workers made to the veterans who served our country. He was proud to share this motivational goal with his employees, who stood a little taller at the end of their shift each morning when the clean, dry, and shiny hallways greeted their visitors. The Medical Center's culture of caring for its veterans also shone through.

Select the "Right" People

Another generic definition of culture is "what we do." When we want to change "*how* we do," one of the approaches is to change our behavior to exhibit the outcomes desired. This "suit up and show up" concept is how habits change. Culture is an accumulation of habits and other elements already described in this chapter. This approach is a little like the chicken and egg problem. Does the evolving culture encourage the individual to change their behavior, or is the behavior change what evolves the culture? The front-line worker's behavior defines the culture; leadership and management can only influence and shape the culture. So how can the organization lead a culture change?

Treatises on forming and developing effective teams talk about getting the "right people on the bus."[4]

Scholtes, Joiner, and Streibel, in their classic, *The Team Handbook,* describe the involvement of engaged individuals under "The 'Laws' of Organizational Change":

> People don't resist change; they resist being changed. The best way to get people to dig in their heels is to give them an arbitrary mandate to change. If you want their cooperation, you've got to keep them on board for every step of the change. Ask for their opinions. What do they hope will happen? What do they fear? What suggestions can they make to ensure the success of the effort? Communicate regularly about progress and results. Provide

a clear picture of what the future will be like, and answers to the questions "How will work be different?" and "When will the changes be implemented?"[5]

Human-Centered Design and Choosing the Right People

Human-centered design is an approach to problem-solving commonly used in the process, product, service, and system design, management, and engineering frameworks that develop solutions to problems by involving the human perspective and emotion in all steps of the problem-solving process. Developing a human-centered organization and culture requires being mindful of both the business and those who are engaged in it.

In an honest exercise of business mindfulness, the human-centered organization and its stakeholders require the humility and the courage to acknowledge that impacts may be generated in all those multiple dimensions, even if some of them are negative and non-intended.[6]

One approach is to assign roles in building organizational culture, considering the personalities of those involved. A retired National Aeronautics and Space Administration (NASA) Quality Engineer related that before a major team effort was chartered, he and the project champion and sponsor would carefully consider who should be on the team. The Quality Engineer Lean Six Sigma (LSS) Master Black Belt, acting as a team facilitator, would then speak with each of the target members, assessing their knowledge, skills, and abilities (KSAs) relative to the outcomes desired. The facilitator met with the individuals long enough to get some idea of their personalities and how their KSAs might contribute, given their personalities and working styles. If additional skills were required, the facilitator arranged for that training for the individual before the team convened.[7]

The Big Five personality test has been used by psychologists for years to measure the strength of certain personality characteristics, the accuracy of which has led to its use by prospective employers and recruitment companies. The idea is that the candidate will have their personality type assessed based on five main characteristics, which are individually scored, resulting in a better understanding of the individual's personality. This then gives prospective employers and recruitment consultants an idea of the employment roles that will be better suited to the individual and allows insight into whether a candidate will possess qualities desirable for a specific role. For example, performance in a career such as "nursing" may be enhanced by friendliness and a caring nature, in which case a high score for "openness" and "agreeableness" from taking the Big Five personality test could indicate job suitability to potential employers. Numerous psychological studies support the correlation between job performance and personality, with some suggesting that almost 30% of differences in performance ability are related to personality characteristics.

The Big Five personality test consists of five personality dimensions: Extraversion, Agreeableness, Conscientiousness, Openness to Experience, and Emotional Stability (also called Neuroticism). Psychometrics, the field of psychology that deals with the design, administration, and interpretation of psychological tests and measures, has researched the Big Five extensively. This test has shown a high degree of predictive validity, test–retest reliability, convergence with self-ratings, and ratings by others.

Scientists have found that the Big Five do predict job performance. A study evaluating five occupational groups – professionals, police, managers, sales, and skilled/semi-skilled – found that higher Conscientiousness correlates with better performance in all these groups. Extraversion predicted performance for two groups that had especially high social interaction: managers and salespeople. Openness to Experience made everyone better at receiving training. Other personality dimensions also helped predict performance in various areas of work for these different groups.

The Big Five model referenced here is just a model. It is meant to guide us, as leaders, to choose the best individuals for the effort. Roles may change as the culture evolves into a positive, uplifting, productive, and rewarding workplace. Organizational leaders must monitor the progress of the journey and adjust the leadership team as appropriate. Like many tools, models are meant to be mixed and matched to the situation.

Once the right people are involved in the cultural journey, it is necessary to continue their learning and expand their ability to lead within the new environment. Other tools can help focus our efforts on the right skills and behaviors to keep our journey going in the right direction.

Empowerment: Nurturing and Developing People and Teams

Human-centered organizations work to motivate and enable their employees to develop and utilize their full potential in support of the organization's overall goals and objectives. Such organizations also work to build and maintain work environments that support their employees and create a climate conducive to performance excellence and personal and organizational growth. People at all levels are the essence of any organization, and empowering them to fully use their abilities and to be fully involved in the organization's processes benefits the organization.

Grace Duffy: *Empowerment means that employees have the authority to make decisions and take actions in their work areas without prior approval, within established boundaries. Years ago, I was working on an improvement project at a hospital in downtown Charleston, South Carolina. The improvement team included a recently graduated nurse assigned to one of the outlying day medical centers, now known as Urgent Care Centers. This young lady gobbled up the quality tools training I provided to the team. She was so enthused at being a part of the team that I and the hospital quality manager asked her to flow chart the intake process for patients coming to her day medical center. We empowered this young nurse and her colleague to manage the process mapping on their own and to present their work at the next team meeting. I was amazed at the growth of this young nurse as she displayed their work. It was as if a small bud had bloomed into a radiant flower. Providing the resources and trusting her to use them to perform a critical step in our project showed this young nurse we truly valued her mind and future contributions to the hospital system.*

Each employee must recognize that the outputs of his or her activities provide the inputs to the next person's process. Chapter 5 uses the SIPOC (Supplier – Input – Process – Output – Customer) and SIPOOC (Supplier – Input – Process – Output – Outcome – Customer) tools to explore the process and system interaction involved. Employee involvement allows employees to participate in decision-making at some level, provides the necessary skills to accomplish the required task, and carefully defines responsibilities and authority. Employee involvement also provides recognition and rewards for accomplishments and enables communication with all levels of the organization's structure.

Managers must do more than just tell employees that they have the authority to participate fully in processes. They must also relinquish some of their authority and show by their actions that they expect full employee involvement and that they support actions taken by employees and decisions made by them to further the organization's goals and objectives. Giving employees the authority to act also gives them responsibility and accountability for what they do. To fully participate, employees must understand the organization's mission, values, and systems.[8]

Grace Duffy: *Empowering employees to perform significant projects is a strong developer of enthusiasm. As head of Corporate Technical Education for IBM Headquarters operations some years ago, I led software applications instructors across Northern New Jersey, Westchester County, New York, and Western Connecticut. Soon after I assumed the leadership position, several instructors came to me recommending that we develop an instructor backup plan. These instructors were willing to provide backup across the whole 110-mile swath from Franklin Lakes, New Jersey, to Southbury, Connecticut, based on their areas of expertise. I was delighted with their initiative and gladly empowered them to develop the backup plan and present it to me. Not only did they provide a complete plan for backing each other up for critical classes, but they also offered a skills gap analysis and suggestions for cross-training new instructors as they were added to our faculty. This gap analysis helped me interview new instructor candidates, including which of the three offices should be their home base. By trusting the lead instructors early in my management tenure, we established an empowering environment where the instructors felt recognized for their competence. I never had to worry about the quality of instruction in our classrooms. The lead instructors managed that on their own and shared the status of activities during our regular review sessions.*

My example of empowered instructors reminds me of a scenario during that same management assignment where I failed to bring the right person into the job. IBM had a full employment policy at the time and when one organization needed to downsize, the company did its best to find equal or better positions for downsized employees in other departments. We were able to place two or three excellent new instructor candidates during this shift in the workforce. One hire I made was not such a good decision. I interviewed a candidate who had the awareness of the need to change her job assignment and the desire to do so. Where I failed to prepare effectively was in assessing her knowledge of the applications she would be teaching and her ability to learn the course curriculum. I still wonder if I had spent more time reinforcing her through the learning process, whether her stay in our department would have been successful. Her colleagues tried to support her slower learning curve until her health started to suffer from the pressure to meet the requirements of the position. Had I been more aware during the interview process, I might have saved her and myself a good bit of frustration and disappointment. Fortunately, we were able to recognize her long tenure with the company with a full retirement package and our thanks for her contributions.

My lesson: Make sure you are empowering the right people.

Help Your Employees through the Stages of Team Development

Most readers will already be acquainted with the Tuckman stages of group development as:

1. Forming
   a. Break the ice, set up for success, set ground rules.
   b. Clarify roles, expectations, charter, and deliverables.
2. Storming
   a. Friction, tension, issues.
   b. Remember that all are normal reactions.
3. Norming
   a. Storming resolved.
   b. Adjustments to roles, rules, styles, and team membership.
4. Performing
   a. Team is excelling, tasks completed, deliverables met.
   b. High accountability.
5. Adjourning
   a. Job is done, and the team disbands.
   b. Document lessons learned and celebrate successes.

The team leader, facilitator, and champion have a ringside seat to watch individuals engaged in the cultural transition as they move through the journey of team forming and growing. How do they deport themselves within the team and in carrying out their roles? What gaps become apparent in their readiness to move to the next step of responsibility and professional growth? The team lead and facilitator continuously monitor team behavior and outcomes to identify opportunities for further support to those engaged in the effort.[9]

The five stages of team development do not have a standard cycle time. A good facilitator, as described in the NASA Quality Engineer example earlier in this chapter, can guide team members through the forming stage with effective project orientation, champion and sponsor introductions, and identification with the project charter. Transparency by the team lead, facilitator, champion, and sponsor about the reason for the project; the desired goal; and how the outcome aligns with core organizational priorities can put the team on the right path to move quickly into later team development stages.

The storming phase must not be ignored. The team lead and facilitator must acknowledge and help the team to work through conflicts or differences in project vision. Often, it is a matter of encouraging open discussion of the pros and cons of the project. A Forcefield Analysis can help team members acknowledge and respect two perspectives on the same issue and work together to agree to move forward. Patience is critical. Often, as a facilitator, it helps to step back and encourage a discussion among equals as the team members resolve the issues on their own. Reinforcement toward alignment with organizational priorities and values guides the discussion toward conflict resolution while identifying issues to be addressed as the team moves into the norming and performing stages.

Kotter Eight-Step Change Model

Kotter's eight stages of organizational transformation are my final recommendation for moving toward a more positive culture. John Kotter, in his 1996 landmark text, *Leading Change*,[10] described an eight-step process (see Table 1.1 and Figure 1.2). This is a universal model to be employed with any visible change effort. Changing the culture cannot be accomplished without the involvement and commitment of the people affected by that culture. Here again is the chicken and egg situation. Does the culture change the people, or do the people change the culture? It is probably a little of both. Once we are on the continuum of change, we will change and be changed by the decisions we make.

Building the team, communicating the vision, empowering action, short-term wins, and continuous reinforcement are characteristics of engagement and trust. A human-centered culture is replete with transparency, trust, and reinforcement.

Culture change may be initiated by senior management, but it becomes a reality through front-line action. Here's a story about a shopkeeper in an older, rundown section of town who decided to do something to clean up his storefront. He started out each day by sweeping the sidewalk in front of the store. He cleaned out his trash barrels and put them out of the way of sidewalk traffic. He made sure his shop window was always clean and repainted the sign over his shop door. He put extra effort into designing pleasing and orderly displays in his shop window. Over time, the other shop owners on the block noticed his efforts and began to clean up their establishments. Because the block looked cleaner, it appeared safer to the

Table 1.1 Actions and Behaviors at Each of the Eight Steps of the Kotter Process for Leading Change

Step	Action	New Behavior
1	Increase urgency	Champions or senior leadership starts telling each other, "Let's go, we need to change this!"
2	Build the guiding team	A group with enough power and respect in the organization are named to guide the change and they start together well.
3	Get the vision right	The project team develops the right vision and strategy for the change effort.
4	Communicating for buy-in	Both technical and administrative personnel begin to buy into the change, and this shows in their behavior.
5	Empower action	More people feel able to act and are motivated to drive toward desired objectives and outcomes.
6	Create short-term wins	Momentum builds as leaders and employees seek to fulfill the vision, while fewer and fewer resist change.
7	Don't let up	The organization makes wave after wave of changes until the vision is fulfilled.
8	Make change stick	New and winning behavior continues despite the pull of tradition, turnover of change leaders, etc.

CREATE
a sense of urgency

INSTITUTE   **BUILD**
change   a guiding coalition

SUSTAIN   *the big*   **FORM**
acceleration   *opportunity*   a strategic vision
and initiatives

GENERATE   **ENLIST**
short–term wins   a volunteer army

ENABLE
action by
removing barriers

Figure 1.2 The Kotter model for successfully addressing "the big opportunity."

public. More customers began to frequent the block and business improved for all the shop owners. One person, embarking on a project to clean up his piece of the block, changed the culture and outlook of the whole street.

The following chapters provide strategic techniques, tools, and ideas to envision, design, and implement a custom-built path to improve organizational culture and move it toward being more human-centered. We have minimized our use of Lean and Six Sigma jargon in writing this book. Those trained in these approaches will recognize many of the tools and techniques when they are described in layman's terms. Start here in Chapter 1, to get to know yourself and your team. Then, progress to choose the most effective techniques to get you to your goal.

Questions for Discussion

1. How do our attitudes and beliefs influence how we engage with others in the workplace? Create a two-column table, with labels at the top of positive attitudes and negative attitudes. Brainstorm one- or two-word descriptions of attitudes that prompt positive behavior or actions in the workplace; list them in the left-hand column. Then, do the same, only brainstorming attitudes that prompt negative behavior or actions; list them in the right-hand column. Note, this is an excellent group exercise.
2. Think of a time when a team member was in a role that was not consistent with their working or behavioral style. What observations did you make about how that person performed in the role? How effective was the person in accomplishing their assignment within that role?
3. Think of an example of a change you went through, either at work or in your personal life. How did you become aware of the change? What impact did the change have upon you? What actions did you take to change your behaviors in response to that change? Organize your example using the Kotter Eight-Step Change Model to describe how you went through the process.

Notes

1. Hacker, Stephen, Foreword to Culture of Quality, by ASQ, Forbes Insight. *Forbes.* New York, 2014.
2. Coll, Josep M., *Buddhist and Taoist Systems Thinking.* Routledge Press, New York, 2022, pp. 74–75.

3. Duffy, Grace L., and Furterer, Sandra L., *The ASQ Certified Quality Improvement Associate,* ASQ Quality Press, Milwaukee, WI, 2020, p. 15.
4. Collins, James C., *Good to Great,* HarperCollins Publishers, New York, 2001.
5. Scholtes, Peter R., Joiner, Brian L., and Streibel, Barbara J., *The TEAM Handbook*, 3rd ed. Oriel Publishers, Madison, WI, 2003, pp. 1–7.
6. Ibid, Coll, p. 183.
7. Adkisson, John. "A NASA Space Coast Kaizen Model." Chapter 11 of Duffy, Grace L. (ed.), *Modular Kaizen: Continuous and Breakthrough Improvement,* Quality Progress, Milwaukee, WI, 2013, pp. 159–182.
8. Duffy, Grace L., and Furterer, Sandra L., *The ASQ Certified Quality Improvement Associate,* ASQ Quality Press, Milwaukee, WI, 2020, p. 17.
9. Ibid, Duffy, pp. 87–89.
10. https://www.kotterinc.com/8-steps-process-for-leading-change/, accessed July 24, 2019.
    https://www.leadershipthoughts.com/kotters-8-step-change-model/, accessed July 27, 2019.
    Kotter, John P. *Leading Change,* HBS Press, Boston, MA, 1996.

Chapter 2

Organizational Alignment

> ...if you don't know where you're going, any road will take you there.
>
> **George Harrison**

Introduction

Every action within the organization should be aligned with some activity that eventually meets a customer's need. Strategic alignment refers to how the business structure fits with the business strategy and its operating environment. When alignment is attained, the firm gains a competitive advantage and increases performance. Alignment is not just for executive levels. To be truly effective, all levels of the organization must tie their activities to key company drivers. Direct line-of-sight from the front line to the board of directors is a significant motivational tool for employee ownership in business outcomes.[1]

Without alignment, there cannot be any improvement made. As we say, "You cannot improve an unstable process." A misaligned organization functions chaotically. To improve it, you need to first stabilize it. Alignment is a key first step to stabilization. Human-centered organization involves employees and leadership aligned on the same path. A full understanding of the goal is critical for enthusiastic engagement in strategic outcomes.

This chapter discusses several perspectives on alignment, from strategic to operational, including the feedback required to maintain awareness

DOI: 10.4324/9781003454892-2

of changes that put pressure on processes and people to alter direction. Considering the human-centered design focus of this text, we address the "Who" of alignment in teams and then the "What" of process and systems management in later chapters. We tie the "Who" and the "What" together in Chapter 11 on Human-Centered Organization.

Essential Alignment

When was the last time you checked your wheel alignment? Most of us know the symptoms of unaligned wheels:

- Tires squeal while turning
- Vehicle pulls to either side while driving
- Steering is not straight while driving on a level surface
- Steering wheel vibration
- Excessive or uneven tire wear
- Poor gas mileage

Grace Duffy: *I remember my first car; a 1960 Mercedes Benz 190B with 4 gears on the column. I loved that car. I paid $500.00 for it. My father drove with me across the country to college in the fall of 1968. It had beautiful red leather seats and a luxurious dark grey exterior with rounded fenders. I named it Nehru since it looked just like the cars the former Indian premier rode in during official parades in the 1960s. My girlfriend and I took the car on a road trip in Utah to visit friends and I hit a pothole with the right front wheel. Thud! Being in the middle of nowhere, we kept going. When we stopped to refuel, the attendant (yes, we had them in the '60s) noticed that the right front tire was threadbare. It cost us almost all the cash we had to buy a new tire, and even more once I got home to have the front end aligned. "Misalignment" was an expensive lesson for me!*

Organizations experience tremendous waste from misalignment. Vague objectives, poorly defined goals, lack of key performance indicators, ineffective communication, or feedback vertically and horizontally across the organization. When executive leadership is not crystal clear about where the company is going, there is no bandwagon for the rest of the workforce to get behind. It becomes a free-for-all.

Aligning a Team

Teams support the execution of organizational objectives at all levels. Team behavior is not only a human characteristic. Figure 2.1 of this chapter illustrates the amazing concert of movement in bird formations called murmuration, but this phenomenon of behavior is not limited to the skies. Aerial videos of antelope or buffalo flowing across the grasslands exhibit the same immediate response to surrounding movement. How does this happen? Individuals constantly monitor the movements of those around them and adjust based on their trust in the leader to take them where they need to go. Is there a version of this group behavior in humans?

Yes. A not-so-complimentary description might be "herd mentality." We prefer to associate this behavior with a highly aligned team. Each member knows the goal of the team and is attuned to the outcome or destination of the whole. Like a murmuration of birds, humans align with others in the team to accomplish stated goals. Birds have their way of communicating direction. Humans communicate in different ways. We set strategies, share these strategies, solicit feedback from the team, and adjust as appropriate, considering environmental and other factors.

We can pause between an event and our reaction to the event. That pause is related to mindfulness, as introduced in Chapter 1. Living mindfully in a state of natural flow is a challenge. It implies adopting a lifestyle based on the premise that less is more, a common premise in Zen (Buddhism) that parallels the concept of Lean Six Sigma. With Lean, less is more. Use

Figure 2.1   This beautiful natural wonder is a "murmuration" – countless birds flying together in an ever-shifting pattern.

only what is necessary to meet the goal. Six Sigma takes the aligned path identified through Lean activities and hones it further by reducing variation to its optimal level.

Covey's concept of abundance is reflected in the Lean approach to the perspective of knowing the process well enough to identify necessities (priorities in Chapter 4) and maximizing outcomes (Y = F(X), in Chapter 5).

Rowing the Boat

Operational-level actions must be consistent with the goals and objectives of the organization. Unless alignment from operations to strategy is maintained, there will be inconsistency in the organizational culture. This cultural disconnect is wasteful. Cultural misalignment severely disrupts communication horizontally and vertically. Think of the crew represented in Figure 2.2. This could be a crew from Princeton University practicing on Carnegie Lake, or from the University of Pennsylvania out on the Schuylkill River in Philadelphia. Every stroke of the oar, when in concert with the others, propels the craft to its goal. If the oars are out of sync, energy is wasted from fighting against different pressures. Organizations are no different. To establish and sustain alignment, leadership must:

- Understand organizational alignment and its importance
- Ensure that every role, entity, and level of the organization has direct line-of-sight to the organizations goals and can align its goals with them
- Describe the importance of having the "right" people on a team
- Establish conditions necessary for team effectiveness
- Identify actions that a team can take when confronted with team challenges
- Execute its own role as a team member

Figure 2.2 Make every stroke count.

Align across All Organizational Structures

Organizational alignment is a systematic, continuous cycle of monitoring core processes to assure that outcomes are consistent with all four segments of the structure illustrated in Figure 2.3. Mission and vision drive the strategy. Process, systems, and structure provide the framework that supports action. People and culture are the foundation, stoking the fire of high achievement.

Each leadership level uses quantitative measures to maintain focus on the overall goals of the organization. When operating results do not match expectations, action is taken to identify the gap, define the disruption, and set an action plan prioritized to the impact of the misalignment. Research has shown that it takes approximately 70% alignment to have any success at all. This level of alignment takes discipline. In one of his best-selling books, *From Good to Great,* Jim Collins pointed out that what high-performing organizations have are disciplined people, who have disciplined thoughts and can carry out disciplined actions.

Figure 2.4 gives a visual concept of the alignment of the corporate vision and goals from executives, managers, and team leaders to the workforce. The workforce assesses overall goals, establishes tasks, actions, and dates for achievement, and provides feedback through management to validate the ability of the organization to achieve desired results.

Corporate strategy indeed comes from top management. That direction will not succeed without the full involvement of the rest of the organization, including partners, suppliers, and customers. A community of action surrounds a successful company.

Figure 2.4 illustrates the top-down flow of policy deployed from executive offices through the organization for translation into operational results. As vision and goals move further into the functional levels of the company, they are translated into measurable objectives and finally into tasks with assigned

Figure 2.3   The four segments of organizational structure.

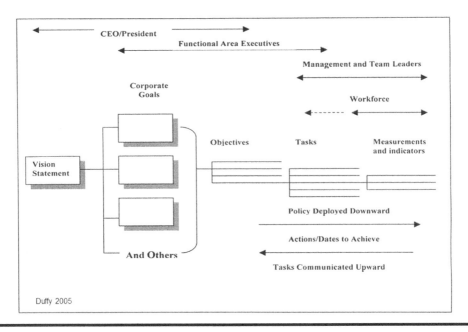

Duffy 2005

Figure 2.4 Alignment "vertically" within the organization.[2]

accountability and verifiable measures. This translation is done best at the level where the action is performed. That is where the intimate knowledge of what it takes to get the job done resides.

Project plans are created at the operational team level, and presented to higher levels of management, as reflected in Figure 2.4 by the arrow labeled "Action/Dates to Achieve." Effective strategic and tactical planning is rarely a single-cycle process. Usually, it is an iterative series of communication from the top down for review, verification, and suggested modifications at the operational level. Once the operations level is clear on their ability to perform effectively against the goals of top management, the project plans are rolled back up to top management for finalization, reporting, and tracking.[3]

Think of the organization as one continuous cycle of human-centered alignment. Encourage your leaders and workforce to:

- Engage! Don't stay on the sidelines, players. Get in the game!
- Act as though every little thing we do matters because it does!
- Don't underestimate your contribution.
- Don't underestimate the power of your attitude. If you are the person everyone is looking at the most – for guidance, inspiration, or direction – a positive outlook is critical.

■ Be FANATIC about being EXCELLENT
■ Get everyone to "pull" together in the same direction

When alignment occurs, everyone wins!

Teamwork

Alignment is only effective when the organization communicates as a series of interconnected teams. Like the rowers from Figure 2.2, we all must row in the same direction, with coordinated strokes.

A team is a group of people who work together to achieve a common goal. Some team characteristics are:

■ Clear purpose. Without a clear direction, there is no team.
■ Mutual reliance. There is a need for each other's experience, ability, and commitment.
■ Clear designation of duties. The whole team knows who is responsible for what.
■ The same definitions. Within the team's process, everyone defines terms the same way. Everyone knows what success, satisfaction, completion, or contribution needs to look like.
■ Belief in the team. The members know that working cooperatively leads to more effective output than working alone.
■ Strong communication. Members document artifacts, manage meetings, and keep one another informed.
■ Shared accountability. The team is accountable to the larger organization.
■ Shared values. The team members share the same values about how they work together.

Remember the statement: **T**ogether **E**veryone **A**chieves **M**ore

Team Effectiveness Model (GRPI Model)

Introduced in 1972 by Richard Beckhard at MIT,[4] the GRPA model is an approach used to describe the four conditions for team effectiveness. The formula is used for leading high-performance teams. It can identify potential causes of team dysfunction and raise awareness about performance issues within a team.

Figure 2.5   Beckhard's GRPI model leads to team effectiveness.

"GRPI" in Figure 2.5 stands for the following conditions:

Goal Clarity: "Where are we going?" With goal clarity, the team
understands what the outcome is; that is, what defines success. Their
bound work is aligned toward a common outcome.

Role Clarity: "Who does what?" This condition is the allocation of
role responsibilities in the work process. Team members know to
whom they should report. Expectations from team members are
explicit. Work doesn't fall through the cracks. If at any point, roles
are unclear, team members should clarify what is expected from
everyone.

Processes, Procedures: "How do we do it?" In this condition, the decision-
making, conflict management, problem-solving, and communication
are delineated. Everyone can apply their limited time more efficiently.
Downstream confusion and conflict are prevented, because team
members know what to expect, and how to get information. Team
members are prepared to do their parts to keep things moving forward.
However, if processes or procedures are vague, team members should
seek clarification, and they should speak up when a process is not
working well.

Interpersonal Interactions: "How shall we behave?" This condition
encompasses agreed-upon guidelines to prevent/minimize confusion
and disruption. Expectations for behavior are clarified. Team
members have a way to address their concerns. Ground rules may
be established to govern team behavior. The top management
structure of the organization can create a culture of quality by
including quality principles in the organization's strategic direction,
and decision-making, and aligning them with other operational
priorities.

Moving from Good to Great

Jim Collins made "Good to Great" a visionary statement in his book of the same title published in 2001.[5] Individuals contributing to an organizational culture of excellence are helping the company move from good to great. Figure 2.6 provides a matrix describing different working attributes that either contribute to or detract from the path toward a culture of excellence. The matrix looks at the combination of Attitude versus Commitment of the individual working in an organization. Alignment with the goals of the organization, both in process and understanding, is a strong energizer for individual commitment. The most effective individual shows the characteristics of a high attitude and high commitment. The quadrants in Figure 2.6 are described below:

Players (high energy, positive attitude)
These members have positive attitudes toward their jobs and the organization and actively invest considerable energy – not just in doing things well, but in making things better. Players believe they can make a difference in the organization, and often do. They may be quite realistic about organizational problems, but also have an optimistic view of the organization's ability to improve. They form the most likely group to support change, both verbally and through action.

Spectators (low energy, positive attitude)
Spectators are likely to speak positively of their jobs and the organization – but seldom make an extra effort or take action that involves doing something new or different unless they are sure it is "safe." They may be introverted, or naturally inclined to "watch and wait" – such people can be pleasant

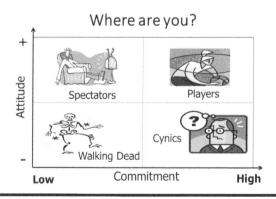

Figure 2.6   Attitude versus Commitment matrix.

surprises when they come out of their shells. Or these people may have had other life experiences that made them leery of putting themselves out there, only to be ignored, or worse, criticized. Imagine the value of energizing this quiet group!

Walking Dead (low energy, negative attitude)

The walking dead have lost interest, burned out, and gotten a bit dull around the edge. They seldom are willing to go the extra mile. They are skeptical about change, stay well within in their comfort zone, and may only do as much as is required to qualify for their paychecks – or avoid being scolded. The walking dead seldom begin that way; once they were Players or Spectators, but they received confusing, demoralizing, or inconsistent messages about the company goals and values. They don't think their work or opinions matter much. They are thinking about the end of the day and the weekend and are not mindful of what is happening because their work environment is unpleasant and unfulfilling. If an organization is heavily populated with the "walking dead," there is a significant cultural problem.

Cynics (high energy, negative attitude)

Cynics often spend a good deal of their time complaining about their jobs, management, policies, or other aspects of the organization. Cynics may be perceived as having a bad attitude; they may even be considered "toxic" in the work environment, as they are squeaky wheels who may spread their discontent and pessimism. They are quick to poke holes in any plans for improvement, predict disaster, and convince others that such efforts are fruitless. However, they are not necessarily lost causes. Cynicism itself is a method of self-defense. Cynics surround themselves with pessimism to protect themselves from anger and disappointment. Mindfulness is completely counterintuitive to cynicism: in a culture where gratitude, positivity, and diversity are embraced and celebrated, confidence improves and cynicism fades.

Traction: The Path to Alignment and Commitment

Misalignment is a common symptom of poor culture, which leads to confusion about roles and duties. When there's a lack of alignment among leadership, management, and frontline employees, important questions are left unanswered: Who IS responsible for that project? Who is leading this initiative? Why are two people working on the same task? When no one can answer those questions, employees are unsure of what should be on their plate. That means overall productivity will lag. All the above problems have a common thread: They limit employees from reaching their full potential at work, signaling the need for a cultural shift.[6]

Grace Duffy: *Some years ago, I had the opportunity to work with an electrical design firm that provided services to the electric utility industry. This small team of talented engineers from Florida Power and Light started their own company, partially out of frustration at the lack of agility exhibited by their previous, larger employer. One of the owners approached me to serve as their corporate quality manager. Executive management had recently contracted with author Gino Wickman to help them implement an operating model aligning all actions directly to corporate strategic objectives. The corporation was growing at light speed and was on the cusp of emerging from a small to medium-sized business, exactly when their business model had to change from entrepreneurial leadership to empowered levels of management. At 130 employees, it was no longer effective to run the business as a skunkworks.*

Wickman's model of 6 focus areas is easily recognizable:

- *Vision*: Each employee knows the leader's view of where the company is going.
- *Data*: Accurate quantitative information is gathered and used for effective decision-making.
- *Process*: Actions, processes, and systems are defined, documented, and integrated.
- *Traction*: The company is agile and can move quickly in the intended direction.
- *Issues*: Challenges and opportunities are clearly identified and pursued according to priority based on the current situation and long-term goals.
- *People*: The company is human-centered and recognizes the value of individual performance.

What makes the model effective is Wickman's concept of "traction" that ties strategic and operational alignment to measurable action. As Wickman states in his introduction to the concept, "In the end, the most successful business leaders are the ones with traction. They execute well, and they know how to bring focus, accountability, and discipline to their organization." Goal alignment keeps all individuals connected to the intended path forward for the company. It minimizes distractions. Literally, traction is where the rubber meets the road. This connection generates a strong workforce and leadership commitment to corporate goals. Each week, every manager meets with their team leaders to focus on what is most important. Nothing is more important than keeping your numbers on track, your challenges (big rocks) on track, and your customers and employees happy.[7]

The agenda for the weekly meeting, named a "Level 10" meeting to indicate the highest level of importance, follows:

- *Segue*: Share good news quickly!
- *Scorecard*: 5–15 most important measurement numbers relative to goals
- *Rock review*: First, the company's big challenges, then individual big challenges
- *Customer/Employee headlines*: No more than 5 minutes for pats on the back, concerns.
- *To-Do list*: Review progress on to-dos from last week, including commitments and accountability.
- *IDS*: Tackle the issues list – Identify, Discuss, Solve. Fast and effective.
- Wrap up and close on time

Grace Duffy: *While I was the contract quality manager for my client, this meeting never took longer than 90 minutes (or about 1 and a half hours). No excuses, other than vacation or death, were accepted for absences. The technical and quality department head led the meeting (in this case, one of the owners) and held each of us, as team leaders, accountable for meeting to-do deadlines. Although the co-owner knew each of us as friends, as well as co-workers, he did not back off holding us directly accountable for a missed deadline, incomplete information, or for being unprepared to discuss a technical challenge or personnel issue. We quickly became a high-performing team, committed to achieving the core goals of the organization within its wheelhouse.*

Summary

Aligning individual workforce goals to organizational goals drives core strategy and priorities.

Aligning all levels of the organization with the mission and vision focuses action on core priorities that meet customer requirements. Empowering teams to achieve outcomes based on performance measures aligned to strategic and operational levels provides human-centered decision-making.

Operating and outcome measures associated with core objectives establish conditions necessary for team effectiveness. True alignment is attained when each level of the organization ties their assigned activities with the required outcomes from strategic goals. The strategic business planning cycle starts with upper management setting goals and sharing them with their direct reports. As the goals are delegated down to divisions and departments, objectives and more specific activities are identified, along with interim and final measures that drive appropriate outcomes. This flow down from top management to frontline worker continues the alignment of shorter-term actions and outputs, including measures to keep outcomes on track to meet customer expectations. Frontline workers provide feedback up toward senior management for review and adjustment as necessary.

Executives, managers, and team leaders are accountable to identify actions that teams can take when confronted with challenges. Team effectiveness and interpersonal skill models, as shared in this chapter, can aid in positioning the workforce for self-reliance and success.

Questions for Discussion

1. What are some causes of organizational misalignment? How do you know when you are not aligned?
2. Discuss how supervisors can help workers clearly identify the worker's role within the organization. How do job descriptions, or employment contracts provide input into role identification?
3. What factors encourage or support your commitment and how can you leverage them?

Notes

1. Duffy, Grace L., *Modular Kaizen: Continuous and Breakthrough Improvement*, Quality Press, Milwaukee, WI, 2013, p. 8.
2. Duffy, Grace L., Original illustration published in Guide to Process Improvement and Change course PPT file, 2005.

3. Duffy, *Modular Kaizen*, p. 29.
4. Raue, Steve, Tang, Suk-Han, Weiland, Christian, and Wenzlik, Claas. *The GRPI Model – an Approach for Team Development*. Version 2. Systemic Excellence Group, 2013.
5. Collins, Jim, *Good to Great*, Harper Collins Publishers, New York, NY, 2001.
6. Kriegel, Jessica, Quickly Shift Your Culture in Three Steps, *TD Magazine*, Association for Talent Development, Alexandria, VA, 2022, p. 45.
7. Wickman, Gino, *Traction: Get a Grip on Your Business*, Ben Bella Publishers, Dallas, TX, 2011, pp. 8 and 190.

Chapter 3

Assessment: How Do You Know Where You Are?

> To succeed in business, to reach the top, an individual must know all it is possible to know about that business.
>
> **John Paul Getty**

If you do not know where you are, how do you know which direction to take to get somewhere else – somewhere better? Figure 3.1 illustrates a snapshot assessment of performance. Organizational assessment is a systematic process of collecting and analyzing data to determine its current, historical, or protected status. There are four basic steps to organizational assessment:

1. Describe the current state.
2. Envision the future state.
3. Assess the gap.
4. Set quantifiable goals.

This chapter addresses the best ways to ensure direct feedback to individual or organizational performance so activities can be adjusted and aligned to shared performance goals. To accomplish this objective, we proceed to understand what assessment is, explain the importance of assessing your performance, and illustrate how to assess organizational performance using a format like a balanced scorecard approach.[1]

 DOI: 10.4324/9781003454892-3

Figure 3.1   Know your business performance.

A reliable organizational assessment is not possible until your organization is at a level of maturity where there are defined processes and systems. Unless there is some stability in day-to-day operations, consistent observations cannot be made on how work is done. Figure 3.2 gives a high-level explanation of the 5 levels of organizational maturity. An organization at Level 3 maturity is one where processes and systems are defined, but the use of data is not pervasive. You could have a well-performing organization at Level 3 maturity because processes and systems have

5. Best in Class

4. Managed Continuous Improvement

3. Defined Systems Approach

2. Repeatable Basic Approach

1. Initial

Figure 3.2   A 5-level organizational maturity concept.

Figure 3.3   Assessing progress and outcomes is essential.

rudimentary definitions. You will not get to data-driven decision-making and improvements until you get to levels 4 and 5 maturities.

Assessment happens only when we have strategic goals, behaviors, data, and culture against which to assess. Identify the best assessment characteristics to ensure direct feedback to individual or organization performance, so activities can be adjusted and aligned to shared performance goals. Assessment is not only for processes but also for keeping the right people engaged in those processes. Likewise, assessment can be performed at many levels.

Let us take a familiar example. What happens when you drive without looking at your dashboard and the road? We have no context to gauge where we are and what obstacles are in our path. Figure 3.3 shows a driver with both hands on the controls, a clear view through the windshield, and an instrument console easily visible to show speed, fuel capacity, and other important data by which to assess the current state. Organizations need this same current state information. Key performance indicators provide information on process (speed), capacity (fuel), and outcomes (operating data). This snapshot provides the basis for assessment against the vision and goals leadership set for the organization.

What Happens When an Organization Does Not Assess Itself?

Moving in any direction without knowing where you are is like springing off a diving board that is not rigidly attached to the edge of the pool. You have no idea where you are going to end up, other than deep water. The concept

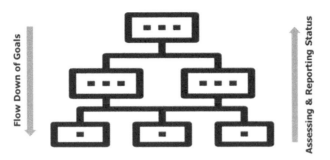

Figure 3.4   Assessment occurs at all levels of the organization.

of strategic planning is based on understanding where the organization is now and where you want it to be at some point in the future. Assessment is a results-oriented review of current performance. It is a key initial step toward achieving performance excellence.

The current state assessment is not just a high-level snapshot of the organization. It is a snapshot of how things are working right now. The only way to know is to look at the results or outcome measures.

Figure 3.4 shows an example of the layered aspect of an organization, and how information flows between levels. Corporate business units flow goals down to their functional units, which distribute subsets of strategic goals to individual departments. Department leaders further decompose these goals into individual assignments with associated responsibilities. Ideally, each individual and supervisor then assesses their performance against the established goals and reports status upward to department heads, who roll the data up to corporate or executive leadership.

This data is then consolidated to show how the reported results relate to the strategic goals initially communicated. This same top-down, bottom-up communication must happen in any organization to stay focused on what will satisfy the customer and sustain operations. As discussed in Chapter 2, Alignment, this top-down, bottom-up cycle provides a human-centered approach to how objectives are met, how they are measured, and how their processes can be improved.

What Is the Purpose of Assessment?

Assessment is the tool we use to identify the current state and compare that with future desired performance. There is external assessment that monitors customer patterns from which we adapt products and services to meet customers' needs. There is internal assessment that provides ongoing information with which to maintain a leadership position within the industry relative to operations effectiveness and efficiency.

Assessments in the form of job and skill analyses are used within the training function for talent management. Both skills training and broader education enhance organizational learning. Leadership and project management skills are needed to accurately orient strategy and tactics to align actions with the organization's values. Scheduled operational assessments keep the organization healthy by identifying opportunities for performance improvement. The world is changing rapidly. Assessment allows us to compare where we are to where we need to be to remain competitive.

We have developed an organizational assessment tool that supports a high-level comparison of a core process to desired outcomes for identifying possible improvements. This tool and a more intensive Best Practice assessment tool are available for all to use. The tools are described in *Validating a Best Practice, A Tool for Improvement and Benchmarking.*[2] *The high-level comparison tool is called the BEST Quick Scan tool. Its use is described in the scenario below and illustrated in* Figure 3.5.

- ■ The BEST Quick Scan tool is an abbreviated 44-criteria format to assess process definitions to see whether enough information is presented for use in improvement activities or benchmarking.
- ■ The Quick Scan is also valuable as an assessment to prioritize internal process improvement opportunities.

Sample Scenario

Grace Duffy: *A new client contacted me to help them prepare for a defense contract audit of their prototype equipment development project. This Skunkworks group of accomplished engineers had finished the customer requirements and design phase of the project and were well through the prototype development phase. I led the executive and senior staff through our Quick Scan assessment tool*[3] *to identify the maturity of a core process. I used the BEST tool Quick Scan assessment (See Figure 3.5) with 42 characteristics based on the Plan, Do, Check, Act model to validate that the core engineering process was aligned with customer requirements and strategic goals. At the end of 20 minutes, we had a gap analysis*[4] *which identified strengths and weaknesses of the design, development, and assembly process.*

		Criteria	NA, C, I
Results		Scope and relevance	
		Integrity of data	
		Segmentation	
		Trends	
		Targets	
		Comparison with benchmarks	
		Cause - Effect	
Enabler	**Plan**	Description	
		Stakeholders	
		Responsibilities	
		KPI's and PI's	
		Deployment and Segmentation	
		Prevention	
		Benchmarking	
		Data	
	Do	Implementation	
		Deployment	
		Cause - Effect	
		Accountability	
		SMART	
	Check	Integration	
		Monitoring	
		Audit	
		Adjustment & Learning	
	Act	Improvement	
		Process	
		Resources	
		Knowledge & Experience	
		Benchmark	
Process		Process description	
		KPI's	
Format		13 criteria	

Code

NA	Not Available
C	Complete
I	Incomplete

Figure 3.5   The BEST Quick Scan Tool.[5]

A second session used the Failure Mode and Effects Analysis (FMEA) tool to prioritize key areas for improvement. The 42 characteristics assessed are designed to drive process maturity to the Best Practice level. Using the assessment rating of each characteristic guided the team through potential failure modes relative to the anticipated contract audit. Areas of improvement centered on documentation, measurements, feedback loops, and communication effectiveness both internally and externally. The FMEA activity identified action items sequenced by priority. Accountability and responsibilities were assigned, with measurements and due dates.

The executive and consultant met with the department heads every two weeks to track progress against assigned due dates. Although the defense contractor chose not to announce their audit visit ahead of time, when they arrived, the engineering and design firm performed strongly against requirements. The contract was not only continued but was expanded because of the maturity of the prototype processes.

Benefits of Assessment

As seen in the previous example, the assessment provided the basis for prioritizing opportunities for improvement. The defense engineering firm quickly identified what was working well in its processes and where the weaknesses were. No time was wasted in rehashing what was going right. Focus was properly placed on areas where most engineers rarely spend time: documentation, measurement feedback loops, and communications. Assessments pinpoint crucial gaps that jump-start a change initiative and allow you to focus your organization on common goals.

What to Assess and When?

Basic business assessments focus on areas such as:

- *Effectiveness*: How well is the process under study meeting customer requirements?
- *Efficiency*: How well is the organization using its resources? Are individuals engaged in the best possible use of their talents? Is waste at a minimum?
- *Satisfaction*: Do all stakeholders view the outcomes of the organization positively?

- *Profitability*: Is the business making a reasonable profit based on industry trends?
- *Return on Investment*: Is net income from operations sufficiently higher than expenses to sustain and grow the business?
- *Benefit–Cost Ratio*: Is the benefit of using resources greater than the cost of the assets?

Human-centered organizations focus on more than mechanics. Yes, satisfaction is on the basic list of assessments. Human-centered organizations broaden their perspective on stakeholders, responsibilities, accountability, adjustment, learning, education, and training. Greater understanding and ownership are developed by engaging employees in the design and implementation of processes and associated training. Empowering employees to develop their own working systems accelerates a culture of excellence. The organization is functioning by doing what they have designed themselves. The system is now "theirs" not just management's.

Figure 3.5, the BEST Quick Scan Tool, is an example of a human-centered assessment, covering not only results and processes but also enablers that are dependent upon communication among internal and external stakeholders. Benchmarking in the "Plan" section of the assessment is a remarkably human-centered activity requiring transparency and trust. Opening the details of how something is done in detail exposes the organization to potential judgment or ridicule. Some organizations choose not to share their process or system details with others, either for intellectual property reasons or simply because they do not want to be embarrassed if another organization can show they function more effectively.

Assessment depends on the goal to be attained. A human-centered organization will assess business, technology, and human elements.

- A short look in the rearview mirror is appropriate to understand where the organization came from and how values were established.
- The current assessment is the basis for a SWOT analysis (Strengths, Weaknesses, Opportunities, and Threats)[6] to prioritize opportunities for improvement.
- Gap analysis can encompass the whole organization or be focused on a single element.
- An ISO audit to a specific standard will drive assessment against the requirements of that standard.

- ■ A human relations assessment will segment toward behavioral and communications aspects.
- ■ A technology assessment will provide current data against which to perform functional or competitive benchmarking.

Looking Only Backward

Can you drive safely by solely looking in your rearview mirror? No. Knowing where you came from is never enough. It is the springboard from which you advance. Culture is established by what people do and have done in the past. It is not static, however. Organizational culture is as important as the core processes that create the service or product a company sells. The holistic system of people, processes, and technology is discussed later in Chapter 11, Human-Centered Organization – A Mindful Culture of Excellence. Assessment of any element within the organization must consider the interaction with all the other elements comprising the organization and the environment in which it operates.

Gathering accurate data about how the organization operates requires the use of both leading and lagging indicators. Leading indicators are process measures taken at critical steps in the service or product creation cycle. These measures are used to appraise current conformance to design specifications. Leading indicators keep the process on track to minimize waste or rework. By measuring key steps within the process, any deviation from expected calibration, performance, or output can be identified before the measurements are out of acceptable range, and action can be taken to return the process to optimal. Leading indicators are generally short-term or operational measures.

Lagging indicators are those taken at the end of a process or major process step. These outcome measures are usually the ones sent to management in status reports. Figure 3.6 indicates the use of leading indicators to measure performance during the activities of a process where lagging indicators measure performance of the finished product or service.

Key Performance Indicators (KPIs) are generally lagging or tactical measures. KPIs are the rearview mirror. They are necessary for reporting and strategic decision-making. They are not the way to drive operations. As indicated in Figure 3.7, the rearview mirror shows us that competition is coming up on our rear fender. It does not help us see what is coming ahead.

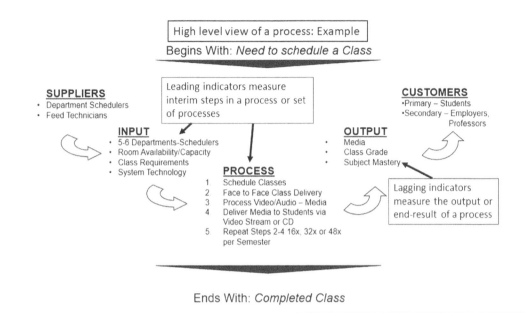

High level view of a process: Example

Begins With: *Need to schedule a Class*

SUPPLIERS
• Department Schedulers
• Feed Technicians

Leading indicators measure interim steps in a process or set of processes

CUSTOMERS
•Primary – Students
•Secondary – Employers, Professors

INPUT
• 5-6 Departments-Schedulers
• Room Availability/Capacity
• Class Requirements
• System Technology

OUTPUT
• Media
• Class Grade
• Subject Mastery

PROCESS
1. Schedule Classes
2. Face to Face Class Delivery
3. Process Video/Audio – Media
4. Deliver Media to Students via Video Stream or CD
5. Repeat Steps 2-4 16x, 32x or 48x per Semester

Lagging indicators measure the output or end-result of a process

Ends With: *Completed Class*

Figure 3.6   Example of leading and lagging indicators in a process for scheduling a college class. Courtesy G. Duffy 2003.

Based on Whose Perspective?

From whose perspective do we assess? Chapter 6 discusses the Voice of the Customer. Customers are just one of the external stakeholders whose perspective we must consider. The eye of the beholder is the target for marketing efforts. When we assess our product and service creation and presentation, we must assess our outcomes as perceived by external stakeholders. Our reputation is an external display of how we are assessed by the market.

Figure 3.7   Driving in the rearview mirror limits our forward motion. Is this a safe way to make your assessment? A real lagging indicator!

Grace Duffy: *Years ago, I was trained as a technical instructor with a large manufacturer of computing hardware and software. My subject at the time was the systems and microcode for complex networking equipment. I remember a senior instructor sharing with me the reminder that although I had taught this code many times and knew it inside and out, the students in my class were seeing it for the first time. My perspective was "this is easy stuff. See, it does this, then this, then this…" The students, my external stakeholders at the time, were wide-eyed with confusion. Their assessment of the code and the class was different from my own. I had to slow down and work at their speed, not mine. I had to assess their skill level and tailor the exercises to novices, not someone who had supported that code in real time for three years.*

Establishing a human-centered organizational culture requires assessing the internal stakeholder continuously. Human resource assessments and surveys abound. All were developed to elicit the perspective of individuals within the organization. Quality assessment tools, such as the Quick Scan tool shown in Figure 3.5, assess the internal process. Skills assessments used by the education and training department provide the gap analysis between current employee skills and those required to meet current and anticipated product and service development. Organizational learning is one of the pillars of creating and sustaining a human-centered culture.

Effective assessments provide answers to two related questions:

1. What things do either external or internal customers consider when evaluating us?
2. How do we know?

We get the answers by direct communication with actual customers, both internal and external!

ASK!!! The Voice of the Customer is addressed in Chapter 6 for external customers, although the same concepts work for internal customers. To attain an internal process perspective, ask those who work closely with a core process, "What internal processes must we excel at?" Ask yourself, what internal processes directly contribute to meeting our strategic goals? What Key Performance Indicators can we set to give us the information to assess the effectiveness of these core processes?

Other processes may not be directly linked to customer-perceived value but provide supporting functions. These indirectly linked processes may be in logistics, purchasing, finance, human resources, marketing, sales, education, and training. These indirect processes, although invisible to the external stakeholder, have an impact on organizational performance and outcomes.

Grace Duffy: *During the H1N1 novel flu virus epidemic of 2010 and 2011, I supported a division of a major U. S. agency responsible for emergency preparedness and response. One of the process improvement projects we used to train agency employees on Lean Six Sigma tools focused on employee travel expense reimbursement. Although this process was not a direct process used by scientists, clinicians, and emergency response teams, having their travel expenses reimbursed correctly and promptly had a direct impact on their ability to perform. An assessment of those involved in international emergency response efforts identified timely reimbursement of expenses as a critical frustration to address.*

The travel office improvement team used flowcharting and value stream mapping to assess the current travel reimbursement process. They interviewed agency workers using the system to identify pain points from the customer's perspective and logged the internal barriers they encountered as they processed expense reimbursement submissions. Three benefits were realized from this assessment. One, the cycle time and accuracy of the process were improved significantly, creating happier customers. Two, the travel office had far fewer frustrations in their daily work, and three, the members of the improvement project came together as a cohesive team, building a positive and dynamic attitude within the department.

Self-Assessment

Assessing your performance helps ensure you are aligned with your organization's priorities. Introspection forces you to take an objective look at your performance. Consider the following questions:

■ What are your biggest priorities right now?
■ Are you on track?
■ Is there anything you should be focusing on?
■ Where do you need to devote more time and energy?
■ How can you help your organization succeed?

Table 3.1 is a worksheet developed as a professional development tool by the head of an economic case for quality committee for the American Society for Quality. This worksheet guides the individual in assessing the

Table 3.1 Self-assessment Form for Personal Alignment with Organizational Goals

Individual Performance Assessment Worksheet:		
This worksheet guides the individual to develop a personally crafted message describing the value they provide to their organization. This value includes, but is not limited to, the knowledge, skills, and abilities gained through professional education and training.		
The flow of this worksheet is based on the concept of Alignment. Use this worksheet to organize your thoughts around the contributions you make as an individual employee or team member to the core mission and requirements of the organization you work for and the specific requirements for which your supervisor is responsible.		
The intent is to show how your activities are linked directly to the core requirements of the company and department. Use quantitative data when at all possible, such as project savings, customer satisfaction surveys or comment cards, production throughput, audit results, service cycle time reports, etc.		
Section 1: Objectives and Key Drivers of the Overall Organization.		
Objectives of the Company:	Category of Organizational Goals or Objectives:	What measurements do the CEO and senior leadership use to assure customer and other stakeholder requirements are met?
Key Drivers:	Customer	
	Operations	
	Financial	
	Learning/ Innovation	
	Other	
Section 2: Department Goals Aligned to Company Goals; these indicators should tie directly to goals in Section 1.		
My Department Contribution to Company Objectives		What indicators does your supervisor use to show the department is meeting its company responsibilities? See the following questions as examples:
Key Driver:	Customer: External	• What is my department's customer satisfaction rating? • How does the company value the outcomes of my department? • What do customer feedback, emails, and comment cards say about the performance of my department?

Table 3.1 Self-assessment Form for Personal Alignment with Organizational Goals (Continued)

	Individual Performance Assessment Worksheet:	
	Customer: Internal	• How is my department viewed by the other departments with which we interface inside the company? • If we work with outside suppliers, what is their opinion of my department?
	Operations	• What are the tangible outputs of my department? • How well are we meeting the demands the company puts on my department? • What is the general opinion of my department within the company?
	Financial	• Does my department stay within our budget allocations? • What does my department do to reduce waste and conserve resources?
	Learning/ Innovation	• What knowledge, skills, or abilities does my department contribute to the company? • Does my department offer additional skills that enhance our customer relationships?
	Other	
Section 3: Individual Contribution to Department Goals.		
My personal contribution to department goals	Department Goals:	What indicators does your supervisor use to track your individual contribution to the goals of the department? See the following questions as examples:
	Customer: External	• What customer satisfaction results, survey comments, emails, etc. provide feedback on your individual performance? • Can you provide activity logs showing direct involvement with external customers?

(Continued)

Table 3.1   Self-assessment Form for Personal Alignment with Organizational Goals *(Continued)*

	Individual Performance Assessment Worksheet:	
	Customer: Internal	• Do you have internal letters complimenting you on work well done? • Is your specific involvement requested by others in the organization? • What documentation do you have of successful team involvement?
	Operations	• What documentation do you have that provides tangible evidence of your contribution to department outcomes? • Can you tie your activities directly to individual department performance measures?
	Financial	• What direct involvement do you have with meeting or exceeding department financial goals? • What documentation do you have to show your actions to reduce waste and maximize the use of department resources? • If your company has a suggestion program, what gains have resulted from your suggestions?
	Learning/ Innovation	• What scheduled or required training have you completed in a timely manner? • What additional training have you completed on your own time to meet department skill needs?
	Other	• What additional activities have you performed that may not directly tie back to Key Drivers for the company, but meet specific department requirements?

alignment of their activities and skills with that of the organization. It begins with identifying the key performance indicators set by the company, based on the Balanced Scorecard model. These are the measures against which the individual and their department must align themselves. The second section asks questions about the alignment between the individual's department and company goals. The third section assesses the individual's contribution to the department and how that aligns with department goals.

This self-assessment can be both a private activity and the basis for a conversation with a supervisor. There are several ways to benefit from the activity.

- Use it as a necessary device for professional development.
- Opportunity to reflect on your career, and not just your job.
- Understand your strengths and weaknesses and how you can improve them.
- Explore how you can contribute more next year.

How to Get the Feedback You Need

Private self-assessment is not the only way to gain feedback for personal improvement or goal alignment. Asking a mentor to support our career or personal progression is another option. Basic steps for engaging with a mentor might be:

1. Understand what you are looking for.
2. Ask for feedback in real time.
3. Pose specific questions and press for specific examples.
4. Turn to colleagues.
5. Use the feedback as a periodic gap analysis/self-assessment for continuous improvement.

Feedback does not have to be a formal event to be helpful.

Summary

This chapter looks at assessment at the organizational and personal levels. Assessment is the tool we use to identify the current state and compare that with future desired performance. It is important to assess current performance at the start of a gap analysis. Without a starting point, it is all but impossible to chart an effective path to the goal. A structured framework helps assess organizational performance.

Table 3.1 provides a model for key performance indicators for all four major components of the organization: customer, operations, financial, and learning and innovation. Self-assessment provides further information which facilitates individual alignment with organizational goals. Removing inconsistencies between personal performance and organizational core processes creates a leaner, more effective product, or service flow.

Questions to ask during the assessment phase are:

■ Do you know and understand your specific goals? How are they aligned, support, and contribute to your organization's goals? How are you assessing your performance against these goals?
■ How are your actions aligned with these goals? Are your actions linked to the appropriate outcomes to support your goals? What are your priorities and are they aligned to optimize your contribution? Chapter 4 will speak about the importance of priority setting.
■ What data would you be using to help with your assessment? A more detailed discussion can be found in Chapter 8 on Data.

Questions for Discussion

1. How does organizational assessment differ from individual or self-assessment? What is the value of each?
2. Describe the process for aligning individual performance activities to the strategic goals of the organization.
3. What is the purpose of measurements and key performance indicators in maintaining process alignment with organizational goals?

Notes

1. The **balanced scorecard** tracks all the important elements of a company's strategy – from continuous improvement and partnerships to teamwork and global scale. Original Harvard Business Review paper, published January 1992.

2. Van Nuland, Yves, and Duffy, Grace L., *Validating a Best Practice, a Tool for Improvement and Benchmarking*, Routledge Publishers, New York, NY, 2021.
3. Ibid.
4. Gap analysis: A technique that compares a company's existing state to its desired state (as expressed by its long-term plans) to help determine what needs to be done to remove or minimize the gap. *The ASQ Certified Quality Improvement Associate Handbook*, Duffy, Grace L., and Furterer, Sandra L. editors. ASQ Quality Press, Milwaukee, WI, 2020, p. 280.
5. ibid, p. 79 f. 3.5
6. Duffy, Grace L., and Furterer, Sandra L., *The ASQ Certified Quality Improvement Associate Handbook*, ASQ Quality Press, Milwaukee, WI, 2020, pp. 204, 205.

Chapter 4

Prioritization

> The key is not to prioritize what is on your schedule, but to schedule your priorities.
>
> **Stephen R. Covey**

If we could go back in time and make one change to improve and benefit our career paths, it would probably be rethinking how we prioritized our daily tasks! Figure 4.1 suggests the importance of focusing on our top priorities.

Time is one of the most valuable and non-renewable resources we have. Unlike other resources, once time has passed, it cannot be recovered or regained. It is important to manage time effectively and efficiently, both at work and in life. Doing the "right" things to achieve your mission each day means prioritizing to meet your major goals. The lesser goals might need to wait a bit, while you attend to more critical issues. Remember the advice of "choosing your battles." You may choose to lose some key battles to eventually win the war.

Figure 4.2 illustrates the sequence of the building blocks of organizational culture. The first building block of organizational culture is alignment, which we describe in Chapter 2. The second building block focuses on the individual engaged in the activity. The "Who," then "What" is so critical to our purpose for this book, that we cover it in Chapter 1. The third building block, Self-assessment, provides a baseline for process improvement. The fourth building block is where prioritization comes in: balancing short-versus long-term goals and improvements to achieve your end goal while optimizing resource utilization. An explanation of the stages is provided in Chapter 12.

 DOI: 10.4324/9781003454892-4

Figure 4.1   Focus on what really matters.

Prioritization is important because it allows individuals and organizations to effectively allocate their resources, such as time, money, and effort, to the tasks and projects that are most critical or have the highest impact. Without prioritization, resources can be wasted on low-value tasks. Important tasks may not receive the attention they need, resulting in missed deadlines, decreased quality, and other negative outcomes. If your organization has well-defined processes and strategic goals, you should at least understand which of those goals and processes is most important. You will focus on the wrong

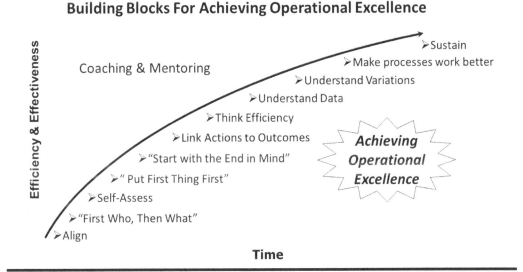

Figure 4.2   The building blocks of organizational culture.

priorities until you know what the "right" things to do are, so don't bother with "doing things right" unless and until you are "doing the right things." By prioritizing tasks and projects, organizations can be certain they work on what is most important, increase their productivity and efficiency, and achieve their goals more effectively. Additionally, prioritization helps to minimize stress and avoid burnout by ensuring that efforts are focused on what truly matters.

Here is a story about a company with its manufacturing line shut down. The owner of the plant was so worried that he immediately called a well-known consultant to come in and help him get the plant up and running again. When the consultant came, he produced a specific tool from his toolbox, then tapped it at various places along the manufacturing line. Doing so, he made his way to the end of the line and tapped one more time. Suddenly, the manufacturing line started running again.

The consultant left and after several days sent back an invoice for $1,000. The owner was shocked. He demanded the consultant itemize the invoice, saying, "You spent five minutes tapping around and charged me this much!?" The consultant acquiesced and sent him an itemized invoice: $1.00 for tapping; $999.00 for knowing where to tap!

The consultant had a deep insight into how problems might occur in a manufacturing line. He could focus his effort on exactly where to fix the problem. As a result, he fixed the problem in an extraordinarily short time and with little effort.

A real-life example that demonstrates the importance of prioritization at work is a project manager who is tasked with leading a team to complete a complex project within a tight deadline. The NASA lead quality facilitator introduced in Chapter 1 used alignment and preventive action to shorten the cycle time for teams to advance from the Forming, through the Storming phase of team development, to get to the Norming or traditional production stage of process improvement.

By understanding exactly how the desired outcome of the project impacted the long-term goals of the organization, the facilitator created a team environment where each member knew their role before the first meeting ever happened. Prioritizing the project relative to other activities within the team members' assignments established the "burning platform" for engagement. The total costs of the project were reduced through effective planning and reduced waste of time and resources.[1] We'll delve further into how teams affect both process outputs and outcomes in Chapter 5.

The experienced project manager realizes that there are many tasks to be done and that the team only has a limited amount of time and resources. If they try to do everything all at once, they will likely become overwhelmed, and the project will not be completed on time.

To effectively manage the project, the project manager decides to prioritize tasks based on their level of urgency and importance. They determine which tasks need to be completed first to keep the project on track, and which tasks can be deferred or delegated to other team members.

By prioritizing tasks in this way, the project manager can ensure that the most important and time-sensitive tasks are completed first and that the team can make the best use of their time and resources. This leads to a more organized and efficient project, with a higher chance of success and on-time completion.

The advice of "prioritizing" is so obvious that one might wonder why we bring it up at all. Clearly, tasks should be sorted by their importance and urgency, and dealt with in a logical order that smooths remaining processes. Is there any other way to operate? Yet if we observe organizations – and individuals – we see repeatedly that knowing the importance of prioritization, and following through with real, useful prioritization, is hardly the same thing. It's rather like how we all know we should eat healthy foods, and exercise regularly. Obviously, we *should* do that, and we understand *why* we should do it…but do we follow through? If so, do we follow through often enough to make a difference?

The key to making the most of prioritization is understanding what actions need to be taken first. This is trickier than it sounds, in complicated situations. "Squeaky wheels get the grease" is a common truth – meaning that the person, thing, or situation that is making the most noise tends to draw our attention. But what if the "squeaky wheel" is merely a symptom of a deeper problem? We can spend hours of our invaluable time chasing after symptoms while solving nothing of the root issue. The root issue remains, and no progress can be made.

How can we get a clearer, practical picture of prioritization?

Time Management Matrix

Stephen Covey's time management matrix (Figure 4.3) is a tool used to prioritize tasks based on their level of urgency and importance. The matrix consists of four quadrants, as follows:

1. **Quadrant 1 (Urgent and Important)**: Tasks in this quadrant require immediate attention and are important for achieving goals and meeting

Figure 4.3   Stephen Covey time management matrix.

deadlines. Examples include crises, problems, unexpected events, deadlines, and high-priority projects.

2. **Quadrant 2 (Important, but Not Urgent)**: Tasks in this quadrant are important for achieving long-term goals, but do not require immediate attention. Examples include personal growth, exercise, learning, and relationship building.

3. **Quadrant 3 (Urgent, but Not Important)**: Tasks in this quadrant are not important for achieving long-term goals but require immediate attention. Examples include low-priority activities, interruptions, and distractions.

4. **Quadrant 4 (Not Urgent and Not Important)**: Tasks in this quadrant are neither important nor urgent and can be deferred, delegated, or eliminated. Examples include time-wasting activities such as watching television or browsing the internet.[2]

This matrix is the basis of the fourth Building Block of Organizational Culture, "Put First Things First" as identified in Figure 4.2.

The goal of the time management matrix is to allocate more time to Quadrant 2 tasks, as they are important for personal and professional growth while minimizing time spent on Quadrants 3 and 4 tasks, which are distractions and timewasters. By prioritizing tasks in this way, individuals can achieve greater balance and focus on their lives and achieve their goals more effectively.

It is important to note that the matrix is not a rigid tool, and regular reviews and adjustments of priorities are essential to effectively use the time management matrix.

The Challenge in Prioritization

As we stated earlier, if prioritization were simple, we'd all be doing it, all day, every day! But there are numerous factors that can disrupt even our best efforts. The challenges we face in prioritizing tasks include:

- **Knowing what really matters**: We have difficulty in determining what is truly important, as priorities can change, and new tasks can arise.
- **Time constraints**: There may not be enough time to complete all the tasks. Prioritization helps to ensure that time is allocated to the most important tasks – but once more, time is limited, and there may be situations in which some priorities simply must be sacrificed or postponed. Knowing what to "sacrifice" can be hard!
- **Distractions and interruptions**: Interruptions and distractions can disrupt focus and make it difficult to prioritize tasks effectively.
- **Emotional attachment to tasks**: People may have an emotional attachment to certain tasks – even identify themselves strongly with certain activities and responsibilities – and find it difficult to prioritize objectively.
- **Resistance to change**: Changing priorities or delegating tasks can be difficult, as it requires people to adapt to new ways of working.

Despite these challenges, prioritization is essential for individuals and organizations to be productive, efficient, and successful in achieving their goals.

Below are some tips for effectively and efficiently managing time and resources:

- *Set clear and S.M.A.R.T goals* (*Specific, Measurable, Attainable, Relevant, Time-Bound*): Establish what you want to achieve and prioritize tasks based on how they support these goals.
  - **Specific**: The goal is well defined in the language the individual understands.
  - **Measurable**: We won't know that we achieved the goal unless we have a quantitative value to compare performance against.

- **Attainable**: The target is set to be ambitious, but realistic.
- **Relevant**: This goes back to our concept of alignment. Does the goal, when achieved, support the outcomes required by the customer and organization?
- **Time-Bound**[3]: Is it clear when the goal is to be achieved? Engagement is difficult when there is no deadline for action.

■ *Create a schedule*: Plan out your day or week, including work, personal, and leisure activities. Block out time for your high-priority tasks first. Be realistic about how much time you have and allocate time for unexpected events.

■ *Prioritize tasks*: Use a time management matrix or another prioritization tool to identify what is most important and allocate time accordingly.

■ *Delegate tasks*: If possible, delegate tasks to others to free up your time for tasks that require your expertise.

■ *Use technology*: Utilize planning tools such as calendars, task managers, and time-tracking software to help manage time and resources. Review your planning tools daily.

■ *Be flexible*: Be open to changes in priorities and adjust your schedule as needed.

■ *Minimize distractions*: Turn off notifications for email, avoid small talk, close your office door, schedule time for face-to-face visits, and/or avoid multi-tasking to reduce distractions and increase focus.

■ *Take breaks*: While it might seem counterproductive, regular breaks can help to refresh your mind and increase productivity.

■ *Learn to say "no"*: It's important to prioritize what is most important and not over-commit yourself, so don't be afraid to say "no" to requests that will take away from your goals or priorities.

■ *Regularly evaluate and adjust*: Regularly evaluate how you are spending your time and adjust as necessary to ensure that you are effectively and efficiently managing your time and resources.

Remember, managing time and resources effectively requires a combination of planning, prioritization, and the ability to adjust as circumstances change. By regularly evaluating and adjusting, you can find a balance that works for you and achieve your goals more efficiently and effectively.

Spinning Our Wheels

Another excellent Stephen Covey parable reinforces the importance of prioritizing.[4]

When faced with a matrix of priorities, starting on the small Quadrant 4 items is easy and can give us a short-term feeling of accomplishment. They go by quickly and we get the reinforcement of checking those fast items off our "to-do" list first thing. We congratulate ourselves on getting tasks out of the way. But wait, those are not the priority tasks that get us closer to our goal. We have forgotten our first building block of alignment. We can spend the whole day cleaning off our desks or returning phone calls and then realize we have not made any progress on what matters. In essence, we have spent hours "spinning our wheels," looking active, busy, and productive even to ourselves, but getting no further down the road.

Covey's parable of the "Rock, Pebble, and Sand" goes something like this:

In this story, a man is carrying a jar and walks along a beach filled with rocks, pebbles, and sand. He starts filling the jar with pretty sand and pebbles, but as he adds more and more of these small items, he realizes there is not enough space for the rocks he wants to save. He then tries to add the rocks, but they do not fit in the jar because the sand and pebbles take up too much space. The man thought for a bit, then emptied the jar into a container, put the rocks he wanted in first, then found that the pebbles and sand had enough room to flow around the big rocks to fill the spaces in between.

The story is often used to teach the importance of prioritization and highlights the concept that if we prioritize the most important things first, we will have more space and resources for the less-important things later. In other words, by focusing on the big rocks, or the most important things, we can achieve our goals and make the best use of our resources.

This story can be applied to many different situations, including work, relationships, personal goals, and more. By prioritizing the most important things first, we can ensure that we are using our time, energy, and resources effectively and efficiently, and achieve the results we want in life.

Mindfulness and Prioritization

When we look at the things that can hijack our efforts toward prioritization, we see that many can be attributed to a lack of mindfulness, a lack of being present, in the moment, in the now. A surprising number of complications ease, if not outright vanish, when we practice mindfulness. When we focus on the present, we ground ourselves in what we can do right now. If this seems vague, let's examine each of the difficulties we discussed pertaining to prioritization, and give each one a "mindful" solution.

■ Knowing what really matters. When we practice mindfulness, we have clarity of thought that lets us see the situation as it really is.
■ Time constraints. So much of our time is wasted by worrying about what needs to happen tomorrow, tonight, or what should have happened yesterday. We are not suggesting that you pretend like there is no future – rather, that you take advantage of what is directly before you.
■ Distractions and interruptions. Mindfulness not only reduces distractions, but forces you to be in the present moment, it also lets you recognize distractions for what they are.
■ Emotional attachment to tasks. Taking a moment to reflect on why you are doing something can be quite a revelation. I knew a secretary who worked two to three hours of overtime each day, certain that it was required to stay on top of an overwhelming workload. When the organization experienced budget cuts and said, "no more overtime," the secretary realized not only that the workload was entirely manageable within an eight-hour day, but that being the "only one who worked late every night" had felt like a badge of honor.
■ Resistance to change. Changing priorities or delegating tasks can be difficult, as it requires people to adapt to new ways of working. Mindfulness keeps us flexible and open, because we can view the situation without anxiety, for what it really is. When we are asked to change course, it is not a personal affront or an attack, merely an adjustment. We remain capable and valuable to the process.

Questions for Discussion

1. Why is prioritization *not* the first of the building blocks of organizational culture? Why do Alignment, People, and Assessment come first?
2. Discuss the value of prioritization and how it facilitates reduced waste and cycle time.
3. Why is Quadrant 2 in the time management matrix the most important one to focus on when targeting organizational long-term goals?

Notes

1. Duffy, Grace L., *Modular Kaizen: Continuous and Breakthrough Improvement*, ASQ Quality Press, Milwaukee, WI, 2013, pp. 159–160.
2. Covey, Stephen R., *The 7 Habits of Highly Effective People*, Fireside Simon and Schuster, New York, NY, 1989, p. 151.
3. Van Nuland, Yves, and Duffy, Grace L., *Validating a Best Practice, A Tool for Improvement and Benchmarking*, Routledge Productivity Press Book, New York, NY, 2021, p. 50.
4. Covey, Stephen R., *The 8th Habit: From Effectiveness to Greatness (The Covey Habits Series)*, Free Press Simon and Schuster, New York, NY, 2004.

Chapter 5

Lean Six Sigma Problem-Solving Framework

> We cannot solve our problems with the same level of thinking that
> created them.

Albert Einstein

As alluded to earlier, there is no need to implement Lean Six Sigma
until you have a stable system. Generally, when you have a misaligned
organization, you have processes that are not repeatable or not well-
defined. Consequently, there's little benefit to implementing something
like Six Sigma since a lot of effort will be wasted. The first four
chapters get us to the point where we have an aligned organization,
and the people within it have clear roles and responsibilities. Their
performances are assessable. Now we are ready to take the organization
to the next level with an aligned team and a clear objective to help
develop the most effective solution for any problem at hand (see
Figure 5.1).

In this chapter, we provide an overview of the Lean Six Sigma
framework, discuss the importance of a structured problem-solving
approach, and provide a detailed description of how it may be
used and applied to improve organizational performance. Lean Six
Sigma is a combined method for improving the quality of products and
services.

DOI: 10.4324/9781003454892-5

Figure 5.1   Problem solving is a team sport for most effective solutions.

What Is Quality?

- All non-price attributes of the goods or services that the customer cares about
  - *Examples*: Meeting commitments, accurate paperwork, response time, perceived effectiveness
- Customer's perception of quality is formed over time by every contact with us
- The customers' long-term, cumulative view drives them to do business with us in the future.

An organization's processes must meet customers' specifications to be successful.

Overview of LSS (Lean + Six Sigma)

Lean Six Sigma is not new – it is an accumulation of skills and knowledge. Lean Six Sigma (LSS) methodology integrates two process improvement models: Lean and Six Sigma, methodologies which were found to have compatible and complementary structures. Organizations can use LSS to eliminate waste and reduce delivery time for services to meet the ever-faster response times expected by clients, communities, and government officials. Its viewpoint is that unless all available resources are directly engaged in the fulfillment of a customer's need, something is being wasted. Leaders can use Lean and Six Sigma to meet the needs of the community, working faster, better, and smarter.

Michael George, in Lean Six Sigma for Service,[1] identifies the major areas of emphasis common to the separate disciplines of Lean and Six Sigma that have been combined into the Lean Six Sigma (LSS) methodologies:

- System-wide integration
- Leadership involvement and visibility
- Business process focus
- Voice of the Customer (VoC)-driven
- Change-management oriented
- Project-management dependent

Lean Six Sigma builds on the practical lessons learned from previous eras of operational improvement. Figure 5.2 gives a summary sequence of major quality improvement pioneers; you can see that we have come a long way in understanding these complicated processes. The early quality improvement activities began in Japan with the work of W. Edwards Deming and Joseph Juran in the 1950s. Subsequent development of improvement models occurred in the '60s and '70s with the Toyota Production System (TPS) (also in Japan) spawning a resurgence of quality systems development in the

Figure 5.2   The quality improvement pioneers.

expansive economic years of the 1980s. Japanese TPS migrated to the United States with Just-in-Time (JIT) concepts of Kanban (pull systems) and other precursors of Lean Manufacturing. Total Quality Management (TQM) grew out of a desire to expand the World War II-era quality control and assurance activities throughout the entire organization. Team-based quality approaches, such as quality circles, followed Dr. Kaoru Ishikawa's book, *What is Total Quality Control*,[2] to the West. At the same time, the international business community was embracing the standardization of quality methods by transitioning the United States military production standards to what is now known as the ISO (International Organization for Standards) standards.[3]

What Is Lean?

Lean is a methodology that focuses on the reduction of waste and greater efficiency to solve business problems.

- Waste is anything not necessary to produce the product or service.
- Waste looks at both the physical path for work and the process steps to remove waste.

Waste is found everywhere, at every level of the organization:

- Management systems
- Communications
- Processes
- Operations
- Projects
- Procedures

Why Focus on Waste Reduction?

As illustrated in Figure 5.3, typical companies that have evolved their processes without using good process development techniques have a large amount of waste that can be eliminated by using the lean systems approach.

Wastes in Development

Eliminating waste during process, product, and service development minimizes the risk of long-term waste once the processes are

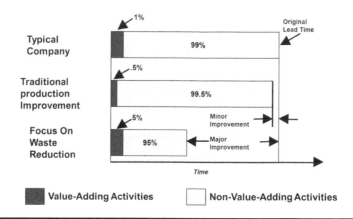

Figure 5.3 Percent of value-adding activities in typical activities.

implemented during production and delivery. The seven wastes identified in Figure 5.4 are the original wastes identified within the Toyota Production Method. An eighth waste, that of wasted human mental capacity, was added later.

There are five essential steps to eliminating waste:

- ▪ Identify which features create value (value is specified from the customer's perspective).
- ▪ Identify the sequence of activities (the value stream).

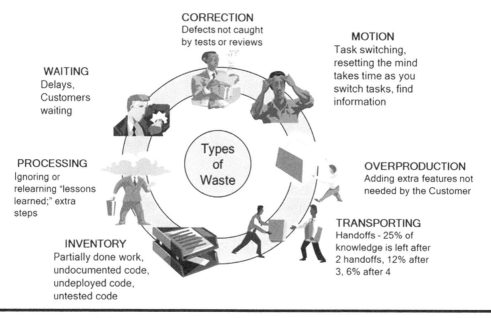

Figure 5.4 The original seven wastes of lean.[4]

- Make the activities or product flow without interruptions.
- Let the customer pull a product or service through the process.
- Continuous improvement in the pursuit of perfection.

What Is Six Sigma?

The concept of Six Sigma was developed in the 1980s at Motorola as an approach to reducing variation in production processes. The primary metric of Six Sigma is the defect per million opportunities (DPMO). Six Sigma is a mathematical term that refers to 3.4 defects per million. In Six Sigma, the higher the sigma level, the better the process output which translates into fewer errors, lower operating costs, lower risks, improved performance, and better use of resources. General Electric followed the Six Sigma path in the 1990s, followed by numerous businesses which embraced the terminology of Six Sigma as their preferred customer-oriented management philosophy.

Six Sigma is a disciplined, data-driven methodology, which links process improvement to organizational strategic objectives, for improving program and business performance. Six Sigma:

- Focuses on process performance by eliminating defects and reducing variation.
- Reduces waste and rework, which lowers costs.
- Creates value through better products and services.
- Establishes a common language and set of tools.
- Identifies what is critical to quality in the eyes of the customer.
- Uses metrics to measure process capability.
- Is about satisfying customer needs economically.

Why Lean Six Sigma?

The need for Lean Six Sigma arises from the need for organizations to stay competitive in today's fast-paced and constantly evolving business environment. With increasing demands for efficiency, quality, and customer satisfaction, Lean Six Sigma provides a structured and systematic approach for organizations to continuously improve their processes and meet these demands. The combination of these two methodologies results in a powerful problem-solving framework that can help organizations to achieve their goals and improve their bottom line. Lean Six Sigma is based on the principles of data-driven decision-making and continuous improvement, which allows organizations to make informed decisions and continuously improve their processes.

Additionally, Lean Six Sigma provides a common language and framework for teams to work together, which can help to break down silos and improve cross-functional collaboration. It also provides a framework for measuring and tracking progress, which can help to ensure that improvements are sustainable over time.

A key phrase in the LSS model is "faster, better, smarter." An earlier phrase, "faster, better, cheaper," was modified to recognize that cheaper is not always the best approach. Wise use of resources to meet critical needs is smart. Reducing costs and expenses beyond the organization's ability to recover from an unexpected occurrence (cheaper) often causes more waste than dedicating the correct amount to begin with.

In transforming organizations to an LSS culture, the three elements of faster, better, smarter are critical for long-term success. Remember, LSS is not a destination. Once started, it is a journey of continuous improvement. If we take our eye off the target or forget to monitor and continuously improve processes, the gains already made will be lost. The goal for leaders is to always seek a faster, better, or smarter way to meet the needs of the community.[5]

Lean Six Sigma Principles for Success

Figure 5.5 illustrates the Lean Six Sigma principles for success. These include:

■ Extreme Customer Focus
■ Process Thinking

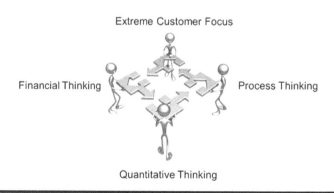

Figure 5.5   Four LSS principles for success.

- Quantitative Thinking
- Financial Thinking

Principle 1 – Extreme Customer Focus

The tenets of Extreme Customer Focus are as follows:

- Quality is defined by the customer.
- Know who the critical customers are.
- Distinguish between requirements given by the customer and customer expectations.
- Flow expectations into the processes that can affect them.

(*Note that this subject will be addressed further in Chapter 6 – The Voice of the Customer.*)

Let's look at an example of this first principle. Figure 5.6 provides an example of the use of Lean Six Sigma improvement methods in a defense contract scenario, to improve the number of satisfactorily closed Requests for Awards (RFAs) and the subsequent increase in award fees. Performance was deteriorating – customer satisfaction trended downward, and the award fees reduced. We had a team meeting with our customers to find out what was not going well and how we can improve our service. After learning about our customer needs, the team went back and implemented various improvements, but to no avail. Our customer satisfaction score kept diving.

We then set about gathering the Voice of the Customer. The improvement team conducted interviews with their client representatives. We brought in some very experienced process improvement practitioners to come back

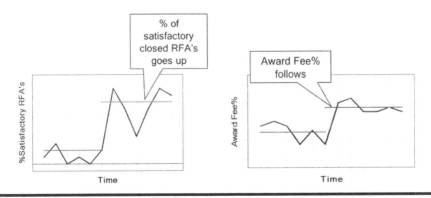

Figure 5.6   Run charts of improvement in closed requests and award fees.

to the table with the customer, to try to understand what drove our low customer satisfaction score. As we conducted the interview, at one point a General, who participated at our interview session, got frustrated and stood up and said, emphatically, "I am a General and I expect to be addressed as General!"

That was when it dawned on us that there were many non-verbal cues that we had failed to consider. Their analysis found that the vendor's Help Desk personnel were not sufficiently trained in communication skills required for speaking with high-ranking military leadership (i.e., military Generals) using their product. Such individuals are accustomed to being addressed with high respect and deference, and their time is extremely valuable. Once we learned that, we implemented a call system that the responder was able to identify the role and title of the caller and greet him/her appropriately by their title! Then the Help Desk personnel were given training and instruction on speaking to high-level representatives of military organizations. As a result, we were able to improve our CSAT score and even received an increased award fee on our program.

Thus, the Help Desk processes were accurate, but the Help Desk personnel were unable to manage the process considering the *customer's* expectations for listening and communication.

Figure 5.6 indicates that additional Help Desk personnel training improved the generals' perception of the vendor's service, contributing to an increase in client satisfaction with the vendor's product and service.

Further Voice of the Customer surveys indicated that this improved Help Desk performance was a positive differentiator for this defense contractor. As satisfaction with their service increased, so did the size of their awarded contracts. Note a process shift occurred in both performance measures. This will be discussed further in Chapter 7, when we look at Process Thinking.

Principle 2 – Process Thinking

The tenets of Process Thinking are:

- All work is a process.
- Manage processes and not outcomes.
- More often problems are the process and not the people.
- Focus on customer value-added activity.
- Improve predictability and stability.

What Is a Process?

A process is a set of interrelated or interacting activities which transform inputs into outputs.

The process elements as depicted in Figure 5.7 are defined below:

- **Suppliers**: the internal/external people or organizations that provide materials, information, or other resources for a process.
- **Inputs**: the resources that are supplied.
- **Process**: the series of work steps that transform inputs into outputs.
- **Outputs**: the product, service, or information that is delivered to the customer.
- **Customers**: the people, organizations, or processes that receive the output. External and Internal Customers.

You can only improve a product/service by improving the process used to generate it. The measures to evaluate process performance will be discussed in Chapter 8 on Understanding Data.

Process improvement efforts are often focused on removing factors from a situation that prevent a process from operating at its normal level. However, much continual improvement involves analyzing a process that may be performing as expected, but where a higher level of performance is desired. A fundamental step to improving a process is to understand how it

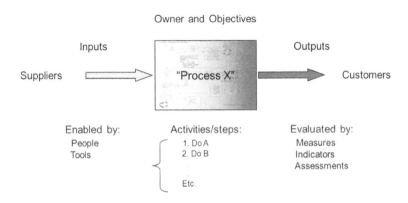

Figure 5.7   A basic SIPOC diagram or suppliers, inputs, process, outputs, customers.

functions from a process management perspective. This can be understood through an analysis of the process to identify the supplier-input-process-output-customer linkages (see Figure 5.7).

The SIPOC diagram provides a foundation for discovery as the process improvement team identifies a standardized process for performing a set of steps to achieve an output.

It begins with defining the process of interest. Suppliers, and what they provide to enable the process (the inputs), are similarly shown on the left side. On the right side, the outputs that the process creates that go to customers, and the customers themselves, are listed.

Once this fundamental process diagram is developed, two additional items can be discussed:

1. measures that can be used to evaluate the performance of the inputs and outputs, and
2. the information and methods necessary to control the process.

This discovery discussion is a basis for a more detailed process map or flow chart showing detailed steps, measures, and data requirements to meet the voice of the customer.

A modification of the SIPOC has been developed, called the SIPOOC (Supplier – Inputs – Process – Outputs – Outcome – Customer), where not only the immediate outputs of the process are identified, but also the system view of the outcome is described. This extension of the traditional SIPOC prompts the improvement team to align the output of the process to the strategic goals or "outcomes" required by their customers.

Principle 3 – Quantitative Thinking – Y = F(X)

Here are the tenets of the Quantitative Thinking:

■ Gather the right data to give visibility into process performance.
■ Look for variation and waste and reduce them.
■ Use quantitative data to drive process improvements.

The third Lean Six Sigma principle for success targets measurements that drive actions to meet specified targets and outcomes. Without measures, there is no tangible alignment with strategic goals.

DMAIC

(DMAIC* is used to explore relationships between "Y" and "X", and to find "F")

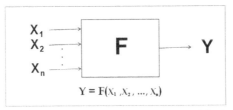

$$Y = F(x_1, x_2, ..., x_n)$$

* DMAIC – Process improvement framework (D-Define; M-Measure; A-Analyze; I-Improve; C-Control)

Figure 5.8   Quantitative Thinking: $Y = F(X)$.

Chapter 8 – Understanding Data covers the basics of how quantitative thinking supports human-centered organizational culture by engaging employees in achievable goals. Without measurable processes, there is no clear line-of-sight on which enthusiastic employees can focus. Quantitative goals provide hard and fast rungs of the ladder that lead to higher levels of performance.

Figure 5.8 represents the concept of $Y = F(X)$ where Y is the required outcome, and X represents the different variables or inputs which the process (F) tasks use to create the outputs that contribute to the outcome (Y). Lean Six Sigma has adopted the Six Sigma process improvement model of Define, Measure, Analyze, Improve, and Control (DMAIC). Although the traditional Plan-Do-Study-Act model described by W. Edwards Deming is an effective model, the DMAIC model calls out measurement as a critical component of process improvement. Without measurement, there is no tangible track record of achievement.

DMAIC: The Foundation of Continuous Improvement

Excellence can be defined as an attribute or a measure against some expectations derived from desired outcomes. As a system, one of the main goals of an organization is to determine what customers (whether internal or external) value, and how to deliver that value most efficiently. The foundation of continuous improvement is based on the Define, Measure, Analyze, Improve, and Control (DMAIC) model.

In the Define Phase, the focus is on understanding the voice of the customer, or VoC. This is Principle 1 of the four LSS Principles. VoC is analogous to starting with the end in mind – the customer's end in mind, as described in Stephen Covey's The 7 Habits of Highly Effective People. Then one needs to listen actively and "seek to understand before being understood," which is another habit in the Seven Habits and the Process Thinking of Principle 2.

Then, the Measure Phase, Principle 3, is about understanding how well things are being delivered to the customer. This is where a baseline measure is gathered to understand the current state. In the Measure Phase, one also seeks to understand the various factors considered as hypotheses that can help explain the level of quality in the outcome.

In the Analyze Phase, one aims to perform a root cause analysis and gather appropriate data to determine the vital few factors that could be improved to yield the most significant impact on the outcome.

The Improve Phase is about brainstorming on potential solutions to address the vital few root causes to improve the outcomes. This is a critical step as organizations only possess limited resources and must focus on the essential few factors to implement the most effective solution for the problem at hand. This effective balance of resources is aligned with Principle 4: Financial Thinking.

Once the solution has been implemented or piloted, we go forward with the process in the Control Phase. This includes any guidance or procedures to the process owner permitting them to monitor the implemented solution and ensure that the process is in control.

DMAIC phases form a scientific approach to problem solving. One develops the hypothesis, gathers appropriate data, and uses that data to prove or disprove your hypothesis so that, in the end, the best solution can be identified and implemented.

Principle 4 – Financial Thinking

Finally, we discuss the tenets of Financial Thinking:

- All process improvement opportunities can be measured in financial terms.
- Value equals quality for the price.
- Select projects based on the value proposition.

The three-legged stool of Lean: Quality, Cost, and Speed remains a pillar of the combined Lean and Six Sigma discipline. Traditional thought taught that only two of the three could be achieved. If quality and reduced cost were achieved, speed of delivery would suffer. If cost and speed were achieved, quality would suffer. Achieving quality and speed would drive up costs.

The systems view of process design, along with understanding the value of effective planning, emerged during the late 1970s and early 1980s with the rise of Total Quality Management. The Toyota Production Method, the precursor to Lean, showed that goal alignment, employee engagement, customer-focused design, and continuous improvement all worked together as a system to achieve all three legs of the lean methodology. A summary of the three-legged stool is as follows:

Link between Quality, Speed, and Cost

- There is a link between delivering a service or product.
  - As quickly as possible
  - With no errors
  - At the lowest possible price
- A process with fewer errors can move at a faster speed (can reduce the variation).
- A faster process is more agile.
- A process that moves slowly may be prone to errors.
  - As products sit around, they may become outdated, damaged, overcome by events, or no longer needed.
- Low quality and slow speed make processes and services more expensive.
  - Work waiting to be completed (such as waiting for a signature to be released) is work that the company has paid for but cannot yet bill.

Mindfulness and Lean Six Sigma

Waste Occurs When Mindfulness Is Absent

The very concept of mindfulness is "lean," when you think about it. Being present in the moment, dealing with things as they come, practicing awareness and appreciation of what is directly before us, and tuning in to the situation are all essential pieces of a mindful state of existence. As soon as we are fully present in the moment, waste decreases. We stop wasting time, wasting energy, and wasting our own mental bandwidth on things that are presently irrelevant.

Let's look at some examples of waste targeted by Lean, as discussed earlier in this chapter, and how mindfulness helps solve them.

Overproduction: When we produce more products than we need, we are overproducing, resulting in depleted resources, increased storage costs, and a lot of trapped revenue. The mindful solution is "just-in-time" production – we maintain full awareness of both demand and supply, and create only what is needed, when it is needed.

Motion: Wasted motion, meaning all unnecessary action, eats up energy and time. Mindfulness gets to work immediately on wasted motion, because we see a clearer path to a goal. We are no longer on autopilot, repeating actions just because that's the way it's always been done before. We recognize when an action is futile or repetitive. We ask questions. "Is there a better way?"

Defects: Defect waste happens when specifications aren't met, and the product does not meet customer's expectations – and it breeds further waste. Being mindful, or fully present, while doing any task greatly reduces the likelihood of mistakes being made.

Overprocessing: This waste happens when we give the customer more than is needed. Mindfulness makes us better listeners, so we can hear not only what the customer wants, but we can forecast also what the customer actually needs based on our insights and on past product demand. Wants and needs are not always exactly the same.

The untapped potential of employees: This waste is a little harder to pin down, but what it means is that an organization fails to make use of its greatest resource: the people who work within it. Mindfulness makes us more aware of the people around us, more respectful of their ideas and skills, and allows us to see opportunity when it presents itself. When people are overwhelmed, they don't have space or time to do anything "right." When they are overloaded, they tend to multi-task everything, and we all know that this is most inefficient because of context-switching. This is partly management's fault for not prioritizing goals. But this is also partly the employee's fault for not helping management and letting them know that there is no capacity left. In this situation, if you need to add anything, something else must give.

Organizational Mindfulness

Listening to anecdotes regarding Lean Six Sigma implementation, we repeatedly hear stories of how extremely simple solutions solve major problems in organizations. In Lean Six Sigma, these solutions are sometimes called the "low hanging fruit" – minimal efforts that produce high-impact results.

Low hanging fruit are incidences of waste that are solved by the implementation not of company-wide overhauls, intensive process changes, or months of measuring and remeasuring, but by a moment of mindfulness, when the leaders and teams experience sudden awareness of the real source of a problem and a comparatively simple way to rectify it. Sometimes, it takes mindfulness to see a situation as it truly is, rather than how we expect it to be, or how we've been convinced by others that it may be. Mindfulness improves our chances of finding root causes, whether they are mechanical or psychological.

Organizations without a tendency toward mindfulness can spend a great deal of money and time tinkering with processes and solutions when, in truth, they have failed to narrow the scope of a project down to the correct scale.

Individual Mindfulness

Individual mindfulness is another way to improve Lean Six Sigma practices throughout an organization, in several ways:

- People who are mindful while they are working are calmer and have better focus.
- Mindful managers have a better grasp on how a process is currently operating. They are present and aware, sensitive to emerging difficulties.
- Mindful employees set goals that are aligned with their organization and are present to focus on their current tasks at hand in support of their goals.
- Mindful workers focus on one task at a time, expending less energy on organizing their thoughts.
- Mindful people hear the real voice of the customer because they are better listeners.
- Mindful people have greater respect for their fellow workers.

Mindfulness throughout an organization, from the top-down and the bottom-up, improves processes at every level.

Simplicity in Presenting Lean Six Sigma Concepts

Hung Le: *The concepts of Lean Six Sigma are not as complex as they may seem, especially when we take the time to break them down into smaller pieces and simplify. Here is an amusing example of this concept from my own career.*

My family and I took a vacation to Hawaii quite a few years ago, during which I was invited to speak with the leadership team at one of the largest building materials and supplies companies in the state. As my 8-year-old son Sean was with me for the trip, I brought him along to the meeting.

At one point during the presentation, I jokingly turned to my son and asked him to share a story about two woodcutters who were working in a forest. Without hesitation and to my surprise, and to the amazement of the leadership team, Sean stood up and shared the story. The first woodcutter was so focused on the task at hand and never took any breaks, while the second woodcutter took a break every so often.

Toward the end of the day, the second woodcutter finished his job a lot sooner to the amazement of the first woodcutter, who then asked: "I have been working so hard without taking breaks! How could you finish your job much faster than I could?!" The second woodcutter responded: "You have not noticed it. But every time I take a break, I sharpen my saw." The story is a clear and powerful reminder that renewing oneself and "sharpening the saw" allows one to be more efficient and productive. This is essential for achieving success in all areas of life.

That was when it dawned on me. This was the same story in Stephen Covey's The 7 Habits of Highly Effective People (the 7th habit), that I told to an audience at another one of my talks when my son, on an off-school day, was on the "sideline" waiting for me. As young as he had been at the time, nevertheless he remembered and understood the meaning and message of the story.

The key lesson for me here was that to connect with people and teach sometimes very complex concepts, you must break it down into smaller pieces, simplifying as much as possible while relate to familiar concepts, like what people see and do every day. It is not much different from a parable or a fable – a story that helps us remember a basic bit of wisdom and easily grasp the key points.

Questions for Discussion

1. Discuss the different process improvement strengths that the concepts of Lean and Six Sigma bring to the combined system of Lean and Six Sigma together.
2. Choose two of the seven original Lean Wastes described in the chapter. How would reducing or eliminating these wastes improve the ability of the organization to meet customer requirements?
3. Describe how having tangible measures of process performance creates a motivating environment for employee engagement.

Notes

1. M. L. George, *Lean Six Sigma for Service*, McGraw Hill, New York, 2003, pp. 6–9.
2. Ishikawa, Kaoru, *What is Total Quality Control? The Japanese Way*, Prentice Hall, Englewood Cliffs, NJ, 1985.
3. ISO (International Organization for Standardization) is an independent, non-governmental international organization with a membership of 167 national standards bodies. Through its members, it brings together experts to share knowledge and develop voluntary, consensus-based, market-relevant International Standards that support innovation and provide solutions to global challenges.
4. Poppendeck, Tom, and Mary, Lean, *Software Development: An Agile Toolset*, Addison Wesley, Boston, MA, 2003.
5. Duffy, Grace L., and Furterer, Sandra L., *The ASQ Certified Quality Improvement Associate Handbook*, 4th ed., ASQ Quality Press, Milwaukee, WI, 2020, pp. 24–25.

Chapter 6

Extreme Customer Focus: VoC (= Y)

> You can't just ask customers what they want and then try to give that to them. By the time you get it built, they'll want something new.
>
> **Steve Jobs**

Customers are rarely concerned with the processes we use to produce what they buy or experience. As indicated in Figure 6.1, customers are interested in whether what they get from us meets their needs and desires. And, as Steve Jobs stated, they don't always know what they want. It is up to us to research, analyze our market, and anticipate what will meet or exceed their expectations.

Once the organization and people are aligned, then it is time to target potential markets. The input of the external and internal customers (stakeholders) is the focus of Voice of the Customer (VoC). Processes and technology are only vehicles to attain the goal of satisfying human beings who experience the product or service. Engaged, enthusiastic employees and leaders use both internal and external VoC input for balancing resources and priorities. This is an iterative process of sustainability which will be revisited at the end of this text.

VoC is an organization's efforts to understand the customers' needs and expectations ("voice") and to provide products and services that truly meet such needs and expectations.[1]

Building a human-centered organizational culture is not a short-term project. That culture influences and is influenced by the stakeholders

DOI: 10.4324/9781003454892-6

Figure 6.1   Time to listen....

with which the organization interacts. Customers are a major segment of those stakeholders. As Stephen R. Covey wrote, "begin with the end in mind."[2] And as Lean Six Sigma states: Y is a function of X (Y = F(X)). Figure 5.8, Quantitative Thinking: Y = F(X), introduces this equation where Y is the required outcome, and X represents the different variables or inputs which the process (F) tasks use to create the outputs that contribute to the outcome (Y). The output of a process is dependent upon the inputs.

The human-centered organization is a long game. Working to meet and exceed the VoC involves gathering data to understand what output of their processes meets the immediate needs or expectations of their customers. However, it does not end there. The human-centered organization is concerned with the outcome of that transaction and how it affects both the external customer and the personnel within the organization. Chapter 11, Human-Centered Organization – A Mindful Culture of Excellence, provides a holistic view of the organization and how mindfulness and empathy are a foundation of a culture of excellence. The systems that maintain an effective organization begin with leadership and strategic vision, then turn outward to focus on the customer. Once the market is identified, we turn inward again to put the right people in the right jobs performing the right processes to create the right products and services.

This chapter explores the difficulties of understanding VoC. We share tools and tips on gathering VoCs and describe some limitations we have discovered about the process.

Objectives/Introduction

Why is it important to gather VoC? We must not only gather the data, but we must also understand it. Knowing the wants and needs of our customers and their perceptions of our quality improves organizational alignment. Unless we know the target, when we are aiming to develop customer loyalty, we cannot set effective goals and objectives. Meeting VoC is a key performance indicator and serves as a base for continuous improvement.

Why Do I Need to Care about VoC?

Customers, both internal and external, seek providers who deliver **value**. Only customers define what quality is and what they are willing to pay for it. Our job is to go to the customer, find out what their perception of quality is, and deliver quality service. As shared in Chapter 4 on prioritization, VoC is a pathway to identifying the highest priority goals of the organization. When running a business, we must ask two questions: "What will satisfy or delight my customers?" and "What must I do to keep the company doors open?" VoC provides the answer to the first question.

What Is Quality?

Quality is a non-price attribute that customers care about. Customers judge quality at "Moments of Truth." In his book, *Moments of Truth*, Jan Carlzon talks about the often-unrecognized myriad opportunities all employees have for gathering customer information. A moment of truth is typically neither positive nor negative in and of itself. It is the outcome that counts.[3] Customers judge our quality relative to whom they view as "doing it best."

"Moments of Truth"

Carlzon, then CEO of Scandinavian Airways (SAS), wrote *Moments of Truth* to describe his approach to customer satisfaction and gathering the Voice of the Customer. SAS was viewed as one of the best airlines of the 1980s when the competition was fierce. Jan identified some of the major moments of truth for his business:

- When and where they use our service
- When they receive a bill for our system
- When they call for help

■ When they read about us in the newspaper?
■ Many other places we often don't know about.

Satisfying a customer's expectations for quality is not always easy. Sometimes customers don't know all their needs, or they state their needs in fuzzy terms. Customers' needs frequently change or their perception of quality migrates. We find customers' perception of quality is ever-changing. Quality is not an event, but a cumulative view. One goof can obviate all the previous good. Without a doubt, different customers have different needs for the same product or service.

Grace Duffy: *I made a huge mistake with a client some years ago. A large power generation company contracted with me to provide a presentation on quality improvement at a corporate conference held on the West Coast of the United States. I worked with the conference planners to identify the topic, their expectations for the session, how many attendees, their demographics, etc. I thought I had properly performed my Voice of the Customer research.*

I flew across the country to corporate headquarters the afternoon before the session and met one more time with the conference organizers. Yes, the content looked wonderful, and the materials were printed and ready for distribution. I was introduced with great fanfare the next morning and began the session. At the first break, a corporate VP came up to me and stridently complained that there was no attendee interaction in the session. I shared that I had discussed the appropriateness of activities with the conference planning team, and it was decided that the venue was not suitable for team activities. The VP violently disagreed and proceeded to take over the session. She announced to all 400 participants that she was dismissing me as the facilitator and would run the rest of the session herself. I and the conference planning team were aghast. Instead of retreating with my tail between my legs, I joined the audience and watched the VP attempt to engage 400 participants in unplanned interactivity.

Hindsight says that I failed to extract enough VoC data. I did not research all the stakeholders for the event. In all honesty, neither had the conference planning team who approached me during the design stage of the event. One of the limitations of VoC is that stakeholders do not always raise their hands to be heard. We need to think carefully about who has an interest in the outputs and outcomes of our products and services. Multiple clusters of customers may have different needs.

One example of confusion in understanding VoC is illustrated in Figure 6.2. Statements such as in the above examples are common, yet they tell us almost nothing about what the customer wants. What are the customer's

Figure 6.2   Voice of the Customer.

commitments? What does "accurate, and right the first time" mean? How does the customer measure "accurate," "on time," or "irresistible value"?

Translate Needs to S.M.A.R.T. Measures

One method to provide clarity to the Voice of the Customer is to apply the S.M.A.R.T. process for setting objectives. In Chapter 4, we used S.M.A.R.T. goals to aid us with prioritization. The same criteria can apply to finding and meeting the VoC. Our S.M.A.R.T. objectives include the following characteristics:

- *Specific*: The target is specified, including the desired result and the reasons for pursuing the objective (i.e., aligning the objective, the VoC goal, and the organization's vision).
- *Measurable*: One or more measures are identified to enable the team to track progress throughout the effort and to know when the objective is achieved.
- *Attainable*: The individuals assigned the objective that can influence or change the process(es) involved.
- *Relevant*: The capability of the process is considered to ensure that the objective is taken seriously and that it is directly related to the Voice of the Customer data. Stretch objectives can benefit customer satisfaction and motivate action, but only if the actions are aligned with specific customer needs or potential delighters.
- *Time-bound*: A realistic date or period is defined within which the objective to be accomplished is acceptable to the person(s) responsible.[4]

What Are the Different Kinds of Needs?

Speaking of identifying customer needs or wants, we also should focus on what customers don't want. Figure 6.3 is a representation of the hidden costs of quality. These hidden costs are often referred to as the Cost of Poor Quality (COPQ) as they are avoidable costs when products and services are created correctly the first time. Employees as well as customers can easily see the tip of the iceberg of rejects, returns, malfunctions, warranty costs, and directly quantifiable defects in our products and services. Each identified quality performance problem carries with it a tangible recovery cost that can be assigned a value. In some cases, however, the value of the intangible costs entailed may transcend the pure economics of the situation. For example, what is the cost of missing an important milestone in a schedule? Quality problems are more often at fault here than other problems. But the most important of all intangible quality costs is the impact of quality problems and schedule delays on the company's performance image in the eyes of its customers.[5]

These "hidden" wastes and costs eat away at our company's reputation. Employees know when they are shipping sub-standard products. Customer service representatives experience added stress when resolving issues that should have been avoided in the design phase. A human-centered culture will pursue VoC input externally and internally. The next section introduces the Kano Model that anticipates customer response to features or benefits of our products.

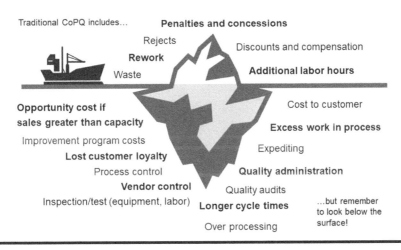

Figure 6.3   Hidden costs of quality and potential customer dissatisfaction.

The Kano Model

One model used to analyze customer satisfaction data is the Kano model (Figure 6.4). Noriaki Kano developed this model to show the relationship among three types of product/service characteristics, or qualities: those that *must be* present, those that are *one-dimensional*, and those that are *delighters*.

The presence or absence of *must-be* or *must-have* characteristics is shown by a curved line in the lower-right quadrant. When a must-be characteristic is not present or is not present in sufficient quantity, dissatisfaction exists. As the characteristic becomes more available or of a higher quality, customer satisfaction increases, but only to a neutral state, represented by the horizontal line. (The characteristic can only serve to not dissatisfy the customer. Its presence will neither satisfy nor delight the customer.)

A *one-dimensional* characteristic drives satisfaction (*satisfier*) in direct correlation to its presence and is represented by a straight line. For example, as the interest rate on a saving account rises, so does satisfaction.

The curved line in the upper left-to-center area represents *delighters*. If absent, there is no effect on satisfaction. But when present, these features delight the customer. As an example, in the early days of the automobile, there were no cup holders. Gradually, auto manufacturers saw the need, and a series of slide-on, clamp-on, and other less-than-satisfactory devices evolved. Eventually, built-in cup holders appeared, and for a time became delighters, resulting in great customer satisfaction. Over time, cup holders became a must-have. Finding cup holders in the new car just purchased is no longer a big deal; not finding cup holders, or not finding enough of them, creates customer dissatisfaction.[6]

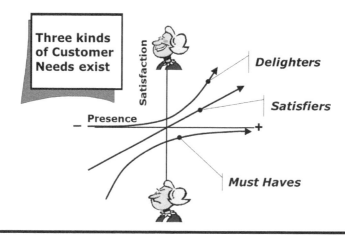

Figure 6.4   The Kano model.

How Do I Gather VoC?

There are two basic types of VoC systems: reactive systems and proactive systems.

Reactive systems are those where customer information comes to you whether you want it or not, e.g., customer complaints. As Tony Robbins, an American author, and coach says, "Leaders anticipate, losers react." Reactive systems wait until they see evidence of a change. Companies that wait to receive a customer complaint before pursuing product or service improvements already have dissatisfied customers.

This is not to say that we should ignore customer complaints. Customers should be allowed to express disappointment or displeasure stemming from a purchase. Complaint data, when appropriately captured and analyzed, provide a wealth of information about customer satisfaction. It is important to recognize that a complaint is not a nuisance; it is a gift. However, we must also acknowledge that this data does not constitute a valid statistical sample: many customers find it a burden to complain unless there is a very serious problem, and many customers appear to have no complaint to register.[7]

What Are Proactive Systems?

Proactive systems are those where we seek out information about customers' needs and wants. Examples of proactive Voice of the Customer approaches are:

- Interviews
- Surveys
- Focus Groups
- Visits
- Benchmarking

This proactive VoC approach signals to the customer and the community that you care about the quality of your service. Customer loyalty is reinforced by each pleasant interchange with a service person, good experience with a product, or productive conversation with a company representative. When the employees of a company honestly care about their customers' needs, they go out of their way to make sure they understand and anticipate customer expectations. This is one of the characteristics of a human-centered culture.

Explore What Is Working Well...

Asking for customer feedback takes courage. We're happy when the customer tells us what works well with the current (system product, service, process) or what not to change. But when we invite customers to share their opinions, we cannot expect constant, glowing feedback.

Grace Duffy: *As an IBM Field Manager in Atlanta, I visited my large system computer software clients every month, hoping to hear that our service was exceptional and that my program support representatives were exceeding the client's expectations. This information is not only deeply satisfying and morale-boosting, but important for continuing to provide excellent results across the board.*

Sometimes the feedback was not so positive. I needed the discipline to listen carefully when a client told me about problems they had with their current service, which is another example of practicing mindfulness. And it's not always easy in a situation of criticism. Our first instinct may be to defend ourselves or our company, make excuses, or search for the flaws in the customer's argument. Fortunately, my employer had effective processes for resolving programming errors or technical failures. When I listened to what the customer expected to restore ideal service, I could then involve the correct specialists to research an issue and achieve a resolution that increased our reputation in the eyes of the customer. The customer and I would set the proper priority for addressing the issue. Some issues were simply an annoyance, and we could provide a workaround until the issue was resolved, usually within a month. Others were such that I would call in the "big guns" and stay at the customer location all night, if necessary, to get the client's systems back up and running. We knew the resources would be there if we used them wisely, so there was little temptation to create a false priority when it was not necessary.

The skill of listening to negative feedback, interpreting (when necessary) the real crux of the problem, and making the customer feel heard takes mindful effort. Though it may seem like a small component to a big process, the ability to listen and then correctly interpret a customer's source of displeasure or dissatisfaction is key. Frustrated customers can be considerably reassured by the knowledge that they are heard and taken seriously, plus it amplifies the chance that the problem is correctly resolved, in a timely way.

Hung Le: *I have a friend who worked for several years in medical malpractice law, and she told me something surprising that has an important lesson to share: "Patients can understand mistakes. They can understand when things go wrong. You'd be amazed how many malpractice lawsuits could be seriously reduced in damages or even avoided altogether if doctors would learn to listen to their patients, apologize when things don't go correctly, and admit that a mistake was made. About half the time, it's when they feel unheard, disrespected or misinformed that patients and their families respond with litigation – and juries never get angrier than when they sense a cover-up."*

Ask About the Ideal Solution...

A useful vehicle for defining an ideal service is a service-level agreement (SLA). An SLA is a commitment between a service provider and a customer. Aspects of the service – quality, availability, responsibilities – are agreed upon between the service provider and the service user.[8] One tool for identifying desired characteristics of service or function is Quality Function Deployment.

Quality Function Deployment

Quality Function Deployment is a system for translating customer requirements into appropriate features at each stage of concept development, from the definition of the function to production, to designing the delivery process, and finally to define the marketing campaign to inform the potential customer of its availability and readiness for use.

The initials QFD stand for:

1. *Q*: The **quality** of your output. How well does it meet and satisfy your customers' requirements?
2. *F*: The **function** defines the size, shape, or form of your output. What do you do or produce?
3. *D*: The **deployment** demonstrates how you do it. Is your deployment process aligned to customer needs and wants?

The QFD process was developed by Y. Akao and T. Fukuhara. The main purpose of QFD is to ensure that the Voice of the Customer (VoC) is captured, analyzed, prioritized, reviewed, and deployed throughout the design or redesign and development process of a product or service. QFD also helps an organization understand how well it satisfies its current customers and what future customers' needs and wants will be for new products or services.

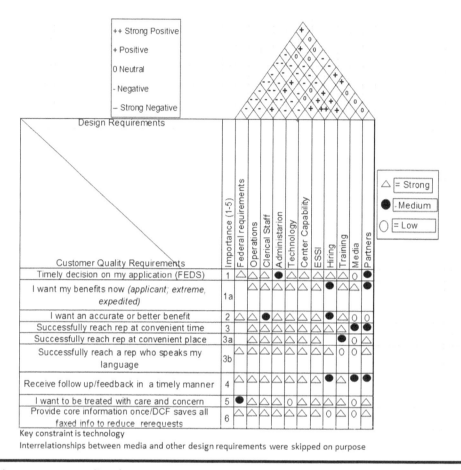

Figure 6.5 Quality function deployment house #1 example.

QFD has a built-in competitive benchmarking process that helps an organization focus on how much it may be necessary to improve not only to meet its customer's expressed needs but to meet and exceed its competitor's strengths. Figure 6.5 is an example of a service oriented QFD activity.

The QFD process is a proactive team-based process that is a structured and disciplined approach to product and service design, redesign, and development. The direct recipient is the client who receives the benefit of experiencing the product or service. QFD supports an organization's initiatives by:

■ Developing an objective definition of product and service quality to be achieved.
■ Teaching the organization about the value of capturing and deploying the voice of the customer throughout the organization.
■ Providing products and services that satisfy internal and external customers.

- Training participants in a tool and technique that can be used in other teaming activities.
- Strengthening the teaming process.
- Helping develop an organization-wide deployment process based on the VoC.[9]

Explore the Future...

Another model that can be used in strategic planning to engage internal customers in VoC generation is a SWOT Analysis. SWOT is a systematic assessment of an organization's internal and external environment. It identifies attributes that affect the organization's ability to achieve its vision and to improve and protect its competitive position. SWOT stands for:

- **Strengths**: internal characteristics or abilities of the organization to meet customer and market needs.
- **Weaknesses**: internal challenges or failings that detract from the organization's ability to perform effectively.
- **Opportunities**: external events or future possibilities that the organization can exploit to become more competitive in a chosen market.
- **Threats**: external events or influences that may negatively impact the organization's abilities to perform effectively in the future.

A SWOT analysis is used during VoC collection in the following ways:

- As a preliminary indication of the current competitive position.
- An assessment of how the organization fits into the current environmental reality.
- Essential during strategic planning data gathering.
- Before starting a major process redesign.
- As an encouragement for minor changes by an empowered team.
- Align future initiatives with outcomes of SWOT analysis as tied to company vision and goals.[10]

About Surveys...

Surveys are another proactive system to collect information about customer needs. Many organizations solicit customer feedback with formal customer surveys. A survey aims to meet as high a response rate as possible to obtain the most representative sampling of the customer population surveyed and as much useful data as possible. Surveys are far more complex than a list of questions to ask customers about a product or service. Designing effective surveys and analyzing the data received are processes involving specialized expertise and knowledge. Combined with Likert scaling, surveys can both capture and prioritize needs.[11]

Mindfulness and the Voice of the Customer

Usually, customers don't want to hurt your feelings or be confrontational. When we ask them questions about our organization, they tend to say what they think we want to hear.

- When a customer says, "I would pay for this!" what they could mean is, "I might pay for this, but if there's another option, I might not."
- When a customer says, "I think your product should have Feature B!" what they could mean is, "I think Feature B would solve a need I have, but maybe there's something else that would solve my need that I haven't thought of yet."
- When a customer says, "I like your product," what they could mean is, "I like your product, but I love someone else's product, and all things being equal, I'd choose theirs."
- When a customer says, "That's a great idea!" what they could mean is, "Hey, it's your money and your company – I don't really care what you do to your product."

These aren't intentional "lies," or even attempts to mislead us. If a person says one thing and does another consciously, we call them a hypocrite. However, if someone does it unconsciously (and we all do so), this is cognitive dissonance. Very simply put, cognitive dissonance occurs when two different parts of our brain are making (or planning) decisions. Customers, like all humans, usually make decisions based on their emotions. And like all humans, they aren't always aware of what their emotions are telling them. What customers want can be driven by processes hidden deep in their emotional subconscious. Therefore, what they *say they will do*, and what they *will do*, are not the same.

Generally, people make decisions in two ways: with their intellect (head) and with their emotions (heart). Your organization may take a lot of time and spend a lot of money to survey customers and listen to focus groups – yet when the time comes for a buying decision, the results go completely against expectations. This happens time and again in political polls – despite asking a million questions to a thousand people, the polls can't predict what will happen. When faced with a set of questions, consumers respond with their heads, intellectually. When it comes time to make a purchase (with their hard-earned money) or make a choice (that will impact their lives), they may go with their hearts.

Mindfulness comes into play in many ways here. First, by recognizing our emotional responses, we become aware of when our emotions are at the helm. "I know an apple is a healthier snack choice, but ice cream reminds me of summers at Grandma's house; that's why ice cream always wins." We are aware that other people often make their decisions on these same impulses and can empathize with their real motivations – comprehending a VoC that even the customers themselves aren't fully conscious of.

Mindfulness also reminds us, within an organization, that the VoC comes from many places. There are multiple touchpoints that we have with customers: our IT department, accounts payable, the help desk, the front desk, the sales department. Empower each employee/customer touchpoint to be mindful and attentive to VoC, encouraging these interactions to be the best they can be. And follow up with these departments as well on their customer interactions. These employees witness real-time customer feedback. Find out what they know!

Conclusion

This chapter shares why VoC is important and helps the reader understand VoC. Once the organization and people are aligned, then it is time to target potential markets. Input from external and internal customers (stakeholders) is the focus of VoC. Processes and technology are only vehicles to attain the goal of satisfying human beings who experience the product or service.

Questions for Discussion

1. Think of a customer service interaction you have had recently. What is positive? Negative? What made it one or the other? What would have made it better?

2. Customers are not all direct purchasers. What other stakeholders should the organization consider when designing a product or service?

3. Internal customers create and are greatly influenced by the company culture. How might executive leadership apply the Voice of the Customer concept to enhance a human-centered culture?

Notes

1. Duffy, Grace L., and Furterer, Sandra L., *The ASQ Certified Quality Improvement Associate Handbook*, 4th ed., ASQ Quality Press, Milwaukee, WI, 2020, p. 313.
2. Covey, Stephen R., *The 7 Habits of Highly Effective People*, Simon and Schuster, New York, NY, 1989.
3. Carlzon, J., *Moments of Truth*, Harper Business, New York, NY, 1987.
4. Manos, Anthony, and Vincent, Chad, *The Lean Handbook*, ASQ Quality Press, Milwaukee, WI, 2012, p. 340.
5. Wood, Douglas C., *Principles of Quality Costs, Financial Measures for Strategic Implementation of Quality Management*, 4th ed. ASQ Quality Press, Milwaukee, WI, 2012, p. 7.
6. Duffy, Grace L., and Furterer, Sandra L., *The ASQ Certified Quality Improvement Associate Handbook*, 4th ed., ASQ Quality Press, Milwaukee, WI, 2020, pp. 241, 242.
7. Ibid. 246.
8. wikipedia.org/wiki/Service-level agreement accessed December 31, 2022.
9. Duffy, Grace L., Moran, John W., and Riley, William J., *Quality Function Deployment and Lean-Six Sigma Applications for Public Health*, ASQ Quality Press, Milwaukee, WI, 2010, pp. 19–21.
10. Moran, John W., and Duffy, Grace L., *Public Health Quality Improvement Encyclopedia*, Public Health Foundation, Washington, DC, 2012, p. 137.
11. Duffy, Grace L., and Furterer, Sandra L., *The ASQ Certified Quality Improvement Associate Handbook*, ASQ Quality Press, Milwaukee, WI, 2020, p. 239.

Chapter 7

Process Thinking

> Worry is the most wasteful thing in the world!
>
> **Unknown**

Process thinking is a systematic approach which drives critical decision-making. At a minimum, this mindful type of planning provides a clear path with various activities that need to be done so that the work can be finished within the expected timeline. In case there are unforeseen events that interfere with the work plan, then the plan can be adjusted accordingly. In the end, one can assess how well the work has been accomplished, and lessons learned can be documented so that the work can be done more effectively and efficiently the next time. Obviously without the end in mind, you can have the best processes and you will miss your goal 100% of the time. This is why it is so critical to have the end in mind and then design the process in a way to get to your goal. With the processes clearly defined, the process can be assessed and improved over time.

How Process Thinking Works

Process thinking (see Figure 7.1) is a way of approaching problems and tasks by breaking them down into smaller, more manageable process elements, then examining the relationships between each element in the process to improve efficiency and effectiveness. This type of thinking is increasingly important in today's fast-paced, constantly changing world where organizations must be able to adapt and evolve quickly to succeed. When

DOI: 10.4324/9781003454892-7

Figure 7.1 Process thinking emphasizes an understanding of systems or events as interconnected dynamic processes.

applied correctly, process thinking can help build a culture of excellence in which everyone is focused on continuous improvement and working together to achieve common goals.

We recall at many of our presentations when we brought up the word "process," people's eyes would glaze over. It seems to be a mental block for many people because in the past they have gone through corporate initiatives where process reengineering was conducted throughout the enterprise. In the end, many of these initiatives failed to yield a sustainable benefit. But that shouldn't be a barrier for any of us. We all depend on processes and the systems they support. Without good processes we would not have a repeatable way to achieve the outcomes we desire. So, it is crucial that we realize and understand that all work is a process. To be able to produce quality outcomes, it's a matter of managing those critical processes. We must manage at the process level, or we will not achieve the desired outcomes. In the old days, like the 70s, when you were responsible for achieving certain outcomes, if you were not able to achieve them, there was a high probability that you would be fired. No one cares to look at the process that is responsible for the desired outcomes. But often, the problems are the process rather than the people. We have seen many situations when the outcomes do not meet expectations, and nothing was done to change the processes so that the desired outcomes could be improved. It seems that people believe that by just checking and checking and focusing on the outcomes that processes will change themselves! This goes with the famous saying: "Doing the same thing and expecting a different outcome is insanity!" Therefore, the only way to improve a product or service is by improving those processes that are used to generate it.

Of course, having processes for the sake of having them is such a wasteful proposition. Over the years, we have observed many organizations whose tendency is to spend a lot of effort to document processes and have them sit on someone else's shelves; and these processes never get used or implemented. So, there really is no value in having them in the first place.

Hung Le: *Some years ago, I spoke with a small clerical pool in a hospital about what I perceived as an overly elaborate process implemented to schedule a department's inpatients for grand rounds and follow-up clinical visits. The process had been worked out by the doctors years before the hospital's patient database was updated.*

When I asked questions about the process, the clerical pool responded, "That way doesn't work anymore; it's just easier to do it this way," and then they showed me how they had altered the process to be more efficient. "Don't the doctors know about this?" I asked them. They actually looked a little alarmed. "No. They don't understand the updated database, but they get angry when anyone tries to change their process." Basically, they confided, the clerical pool achieved the same results much faster, without angering the doctors who believed their own outdated process was infallible. "We just print it out to look like the old reports," said one, "and the doctors never notice the difference."

This astonished me. Basically, I was seeing two groups of people, ostensibly meant to be working together with their overriding goal being providing excellent, efficient patient care. On one side, the doctors were unaware or unwilling to embrace advances in technology, and unwilling to trust the people who worked for them to know the best way to use that technology. On the other side, the clerical pool was forced to make useful, and possibly necessary, changes "on the sly" so they didn't get in trouble, plus they performed an additional task to cover their tracks. Core processes that support strategic outcomes are the life blood of the organization. When both employees and management know and maintain core processes, everyone understands the goal. This alignment with strategic outcomes is a huge motivator for a culture of excellence. Focus on what adds value for the customer. What adds value for the customer also reinforces the company's competitive edge.

Why Process Thinking?

One of the key benefits of process thinking is that it helps organizations to identify inefficiencies and areas for improvement. By breaking down tasks and examining each step, it becomes easier to see where bottlenecks are occurring or where processes could be streamlined. This information can then be used to make changes that will improve efficiency and productivity, which can have a significant impact on the bottom line.

In addition to identifying inefficiencies, process thinking can also help organizations to improve their ability to respond to change. When processes are well-defined and understood, it is easier for everyone to adapt to new situations and implement new solutions quickly and effectively. This is particularly important in today's fast-paced world, where organizations must be able to respond quickly to changes in the market or their industry to stay competitive.

Another important benefit of process thinking is that it promotes a culture of continuous improvement. When everyone is motivated to examine and improve processes, it becomes part of the organization's DNA. This creates a sense of ownership and pride in the work being done, which can lead to increased engagement and motivation among employees.

Finally, process thinking can help build a culture of excellence by fostering collaboration and teamwork. When everyone is working together to improve processes and achieve common goals, it creates a sense of unity and shared purpose. This builds stronger relationships between employees and increases overall morale, which can have a positive impact on the organization's bottom line.

Value-Added vs. Non-value-Added Processes

Figure 7.2 represents all tasks of an organization. Start with core processes to analyze which tasks add value to the product or service. Make note of any task that does not add value. Some non-value tasks, however, are required by law, by industry standards, or customer contract. Those tasks, if nothing else, are done to keep the doors of the company open and must be performed. They are called "non-value-added essentials."

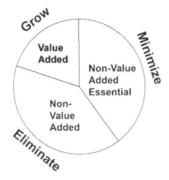

Figure 7.2   Grow value added tasks, minimize non-value, non-essential tasks, eliminate all else.

Any non-value-added tasks should be eliminated.

A good description of non-value add tasks or activities are the 7 Wastes of lean identified in Chapter 5, Lean Six Sigma Problem Solving Framework. Short explanations of value-added and non-value added are listed below.

■ Value Added:
  – Any activity that changes the form, fit or function of a product/ transaction
  – Something the customer is willing to pay for
■ Non-Value Added:
  – All other activities are WASTE (examples: inspection, waiting, transportation, etc.)
■ Non-Value Added Essential:
  – Required activities due to CMMI/ISO compliance, regulations, laws, process performance, and others

Process Mapping and Streamlining the Process

To improve a process, you need to understand how work is performed. This means that one must map out the steps or sequence of steps representing how activities are done currently. So, it should focus on the current state or the as-is process, and not what the process could be or what it should be.

In this book, we will not present an in-depth discussion around how a process map should be completed. We will just highlight the three approaches, then refer the reader to other information sources so that he or she can learn more about these different techniques. Essentially, the three

Basic Flowchart	Activity Flowchart	Deployment Flowchart
• To identify the major steps of the process and where it begins and ends	• To display the complexity and decision points of a process	• To help highlight handoff areas in processes between people or functions
• To illustrate where in the process you will collect data	• To identify rework loops and bottlenecks	• To clarify roles and indicate dependencies

Figure 7.3 Types of flowcharts.

approaches are: basic flow charting, activity flow charting, or deployment for charting. See Figure 7.3 for guidance on when to use each type of flow chart. Generally, as a best practice, the deployment flow chart is most often used because most of the issues relating to process have to do with the handoff areas, where people from different functions are engaged. The deployment flowchart helps clarify roles and responsibilities to ensure flawless handoff between two different groups.

Steps to Streamlining a Process

Streamlining a process focuses on simplifying and optimizing how work is performed to improve efficiency, reduce waste and minimize defects generated from the process. Generally, the steps involved are:

- Identify the process that needs to be streamlined. As part of this step, the goals, the objectives and the scope of the process to be improved are defined.
- Create a visual representation of the process using a mapping technique as described above. This helps to identify any unnecessary activities, redundancies, or bottlenecks in the process.
- Review and analyze the process map to uncover any process issues and identify opportunities to eliminate waste, reduce variation, improve quality, and increase efficiency.
- Based on the above analysis, redesign the process to optimize the process flow. Also, look for opportunities to automate and/or standardize the process as needed. Involve process owners and key stakeholders to get buy-in, as necessary.

- Implement the new process. Communicate the changes, provide training, and ensure that everyone involved in the process understands their roles in the new process and can implement it. Pilot the new process to ensure that it works as intended.
- Monitor the new process to ensure that it is working as intended. Regularly review the process and adjust as needed.
- Continuously review and further improve the process.

Important Things to Remember

Help people see and understand customer value. Workers who understand the "why" of their jobs and know the results of their labor take pride in it, and are better equipped to improve the processes they work in. Don't assume that your workers know exactly why your product or service must be a certain way from the ground up; and don't assume, either, that they can't come up with a better idea.

- Turn over all rocks. Under every rock there is an opportunity to find and eliminate waste
- When in doubt, simplify and eliminate. Focus on value added opportunities.
- Avoid the batch and queue method as it involves producing or processing many items in batches before moving them to the next stage in the process. This often leads to high inventory levels, long lead times and waiting times, and quality problems.

Go and see! Get out of the office, your workspace, your headspace, and find out what your fellow workers are dealing with. This is about communication and empathy. When a process is failing and you only know one small part of it, you cannot contribute to a truly meaningful solution. Look for opportunities to improve and engage your fellow workers.

Fight complacency. When a process gets the job done, we tend to let it ride. The old saying is, "If it ain't broke, don't fix it." But a process that "ain't broke" might be barely working anyway and could be vastly improved. And we do have that tendency to say, "Well, we've always done it this way…" and then relax back into old habits. Change seems difficult, challenging, even scary – yet what could be worse than being left behind, losing customers, revenue, and eventually the organization itself?

Mindfulness is the first step to combat complacency. The moment we can view a situation clearly and without bias, we begin to see where improvements can be made. Empowering mindfulness at every level means a wide-awake work force that sees the process from all angles. An organization can fight complacency by stressing communication, tracking technological updates, watching what competitors are doing (what they are doing better, and what they may struggle with – lessons can be learned from both!), and measuring progress.

Do not focus on punitive measures. When a process isn't performing to expectation, it's common to try blaming employees. But workers who are punished (such as reprimanded, embarrassed, or curtailed) respond with increased stress and decreased morale, leading to absences, higher turnover rates, more mistakes, and motivation to do "just enough" not to be punished again. Worse, employees who are hyper-focused on simply meeting a quota or some other measurement will often do so to the detriment of the process as a whole.

For example, let's say Manager Amy's department produces 100 widgets a day, and 10% of them fail a requirement, preventing their passing on to the next stage. Manager Amy is called out and humiliated each week at the team meeting for her department's poor performance, but she is in no way empowered to examine or improve the process itself. Manager Amy decides that the best way to avoid this is to stop checking each individual widget for that requirement. Now she checks one in ten, and in this statistically perfect example, 99% of her widgets pass on to the next stage. Now she can go to a team meeting and be left in peace.

Down the line, widgets go out and customer satisfaction begins to drop because some of the widgets are defective – but by the time customers complain, and ask for refunds or replacements, the matter is far outside of Amy's purview, and she doesn't care. She just wants to sit through a team meeting without being grilled.

There are two points to note in this example. First, Manager Amy was put in a position of defensiveness which made her look for a solution to her own stressors rather than a solution to the problem itself. Second, examining the process, finding where and why the failure was occurring, and given the ability to rectify it, Manager Amy's department could consistently meet the requirement (a big improvement for both Manager Amy and customer satisfaction) and the need for inspection could cease altogether (eliminating a non-value-added activity).

People are amazingly clever; organizations can use that priceless resource or not. Put in a position of self-defense and self-preservation, employees will adapt with "workarounds" and tailor their results to match a pre-ordained structure, no matter the cost. When employees are motivated by rewards, loyalty and pride, by understanding the total system, by seeing the results of what they make and do, all that amazing cleverness will work for the organization.

Without leadership support, change simply won't happen. The culture of an organization comes from the top – so C-suite executives that want any change in the organization need to start with their own behavior and practices. The organization must be empowered to change, and it cannot be unless that empowerment comes from the top. Those at the top who think they can "talk the talk" without "walking the walk" are fooling themselves.

Mr. Smith's law firm – where his name had been on the front door for more than 30 years – decided to go completely paperless, both to save copy and paper expenses and to be environmentally friendly. The change was planned out well in advance, and with Mr. Smith's blessing. Preparations included employee and attorney training. Plus, a considerable initial expense was incurred to transfer all paper documents to electronic files. The change happened on January 1. But Mr. Smith decided quickly that he didn't want to work electronically after all – he liked having his paper documents. He refused to go paperless – and therefore, the office didn't go paperless either, despite all their efforts. As the boss who wanted his work done on paper, Mr. Smith created a system where paper still had to be used by almost everyone, because they all reported to him. What's more, his attitude trickled down, as senior attorneys, then junior attorneys all realized that the "paperless" office wasn't a requirement. Because they were more accustomed to working with paper, it was just easier to stick with the old ways. When the employees saw Mr. Smith ignoring his own policy, the change initiative was "dead on arrival."

Don't look for "quick fixes" or "one-stop solutions." This is about continuous improvement. We will not find a single solution to fix every problem and make a perfect process. That's impossible. Nor should we make one improvement – even if it's a great one - and consider our job finished. That's complacency. A culture of improvement means we are always looking for opportunities to get better while still appreciating the hard work we do, and the good results we get at present.

Don't play the shell game with waste. Moving waste to another shell, or "sweeping dust under the rug," is NOT improving the entire system. Think holistically. Ensure that vastly changing a process in one area isn't causing increased problems in another.

Using a tool is NOT transforming a culture. This is a common organizational mistake, and a very expensive one! The perception is that "If we spend a lot of money on this X, our organization will improve!" What is X? You name it: a software program, a training program, upgraded equipment, any shiny new thing – and that includes Lean Six Sigma. Tools are only as good as the hands you put them in. Tools do not make workers engaged and mindful. Engaged, mindful workers will tell you what tools they need.

Busy Bees. Increased activity is NOT progress. Ensure your LSS activities add value. Don't assume that adding tasks is the same as adding value. The goal of LSS is not to have everyone "look busy" all the time to impress the CEO when she walks the floor. People busy doing something and people busy doing nothing look very much the same; but the end results are quite different. Mindfulness resists busy-work, and busy-work is notorious for masking real process problems. When workers honestly have nothing to do, this is a sign that there is waste happening in a process and thus, there is an opportunity for improvement.

Questions for Discussion

1. Why is it important that process thinking includes consideration of how the processes fit into the larger system to create not only outputs but the outcome?
2. Discuss the difference between critical (core) processes and supporting processes.
3. How does working "with the end in mind" create a motivating environment for employee engagement?

Chapter 8

Understanding Data

In God, we trust. All others must bring data!

Edward Deming

A Story about Data Usage

Hung Le: *I embarked on a trip along Highway 405 in Northern California a few years ago. At one point, I looked behind me and saw a police car flashing its light. So, I slowed down immediately and pulled over. The police officer politely asked me for my driver's license. After checking it, he asked whether I knew why he had pulled me over. I responded: "I am sorry, Officer. But I am not sure." He then told me that my speed was recorded well over 20 miles of the speed limit per his radar readout. After a short exchange, he wrote me a hefty speeding ticket and informed me of a court date if I chose to dispute it. We then departed on our separate ways.*

The above story shows that I had the same information as the officer – my car's speed. Yet, I did not use it or pay attention to it so that I could stay within the required speed limit. Consequently, I received a speeding ticket!

Data is all around us. To effectively manage business performance (see Figure 8.1), we must harness the "right data as we are constantly faced with overwhelming amounts of data generated and thrown at us. Advanced modern data systems can spew up-to-the-minute data reports and we assume it must all be relevant. It's data, right? It means something. In our present world, we have so many data at hand that it overwhelms many of us – we don't know how to handle or use them, whether they are relevant

DOI: 10.4324/9781003454892-8

Figure 8.1   Using data to manage business performance.

to us, or whether these data are accurate. We are in a world rich in data but poor in information. Most of us were taught to respect data but few of us were ever taught how to interpret it.

We are vulnerable to those who would use data to mislead us. Here's a completely fictional example, but it will probably sound familiar. "An amazing new study shows that 58% of people who contract hepatitis eat broccoli more than once a week," warns the news anchor. We don't want to contract hepatitis! We stop eating broccoli, broccoli sales plummet, and the rate of hepatitis doesn't change one bit, because this piece of data was cherry-picked (or should we say broccoli-picked?) from a study with a limited population, we were given no context for the data itself, and many of us are fuzzy about the difference between correlation (broccoli is eaten twice a week by many people, some of whom will also contract hepatitis) and causation (eating broccoli leads to contracting hepatitis).

Similarly, organizations have access to massive amounts of data but often struggle to turn them into actionable insights that can improve their decision-making process. Quarterly reports are issued that show lists of numbers, which are interpreted with no context and with poor understanding of correlation, sending all levels of management into a panic-mode and resulting in angry meetings and sleepless nights. Or, that same list of data might indicate, at first glance, that everything is fine…but anyone with a keener understanding of what data means can see warning signs

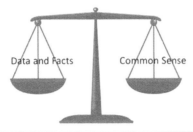

Figure 8.2   Decision-making using data & facts versus common sense.

that, unless some action is taken, the organization will run into trouble in a few months. There may even be a third scenario: reports are generated that provide pages and pages of data that look great but mean little – nowhere in this report is there a real measurement of what truly matters about the organization's performance!

In situations where we have little or no data, we tend to rationalize that it is all common sense or that it is intuitively obvious since we had so much experience or insight about a subject. But as Voltaire pointed out: "Common sense is not so common!" Another wise man also said: "Common sense is the collection of prejudices acquired by age eighteen!" So, one must be very careful when common sense is enacted. It is common sense now that in automobiles, children should ride in age-appropriate, properly-installed car seats to keep them safe – but anyone who is more than 40 years old probably remembers a time when no such rules applied. What comprised common sense back then?

Of course, this is not to say that there is no place or need for common sense to be applied. As Figure 8.2 shows, it is a balancing act. There will be times when data should be gathered to validate assumptions you may have that are deemed "common sense," when they truly are not.

Overcoming this challenge requires organizations to adopt a human-centered culture and then invest in the tools and expertise to turn their data into actionable information. Doing so will allow these human-centered organizations to turn their data into a strategic asset and gain a competitive edge in today's data-driven world.

What Is Data?

Data consists of any information or facts that can be collected, analyzed, and used for making decisions or gaining insights. It can include numbers, text, images, or audio. In effect, it measures an event, phenomenon, or observation. It can be a numerical value, a categorical value, a string of text, or pixel values for images. Examples include data collected through a survey, data collected through sensors, or data collected at various levels of detail, such as daily, weekly, or monthly over different time frames.

Data plays a crucial role in modern organizations, as it provides valuable information that can be used to make informed decisions and address issues and problems. Effective use of data can help organizations improve their operations, increase efficiency, and gain a competitive edge.

In practice, data generally have the following characteristics:

■ Gathered and collected purposefully
■ Communicated and transmitted intentionally
■ Used collectively to make decisions

Why Worry about Measurement Data?

The answer is simple: accurate measurement data tells you if you are achieving your intentions. Without looking at measurement data, we must simply make assumptions. Failing to measure and interpret our situations correctly results in wasting time, money, personnel, and in fact, the reputation of the organization.

Measurement data is the foundation of how we manage the performance of a process. Without data, knowing how well we are doing is hard or even impossible. We may think we are already doing well when we are not, or we may be operating just well enough to get by, never realizing the vast improvements that could be made. We can even waste resources to improve processes that need no improvement. Measurement data lets us know the real story.

Without data to support decisions, management boils down to a set of guesses, operating on gut instinct. We believe the situation is evident, or it is a matter of common sense, so we act accordingly…and in worst-case scenarios, we can make expensive blunders or demoralizing errors. (Stephen Covey said, "Common sense is the least common of the senses!") If there is a process that needs to improve, data measurement lets us identify, describe, and set priorities on the problem. With data measurement, we know what issues should be tackled first (the biggest wastes, for example, or the most critical and serious aspects to finding a solution).

Measurement data also levels the playing field for employees. Employees given clear data can fully understand what is expected, when they are making progress, and when they are working on the right things, and, likewise, when a process is lagging or otherwise failing. Data provides "triggers" that say when it is time to adjust. There is an objective, equitable basis for comparisons (meaning rewards, or in some cases, punishments). Participants have accountability and transparency in the decision-making process.

Finally, without measured customer data, we cannot gain a deeper understanding of customer behaviors, preferences, and needs. We use surveys, focus groups, and interviews to learn more about the voice of the customer.

Importance of the "Right" Data

Having reviewed many process improvement projects in our careers, we have seen projects that collected and analyzed a lot of data. In the end, however, though many graphs and charts were generated, few insights were drawn from such an analysis.

Building a human-centered culture requires that people think about using data to get better insight into what they are working on. This will help them make better decisions. To do so, they need more than just gathering data. They need to gather the right data.

This is critical for an organization to make informed and effective decisions based on real insights rather than assumptions. Having the right data also helps organizations to allocate resources more efficiently by pinpointing areas that require more attention and resources. Customer satisfaction improves with the ability to understand more about the customers and their needs. With the right data, organizations can identify inefficiencies and areas for improvement, leading to optimized processes, reduced waste, and improved productivity. Finally, by analyzing the right data, organizations can gain insights into new trends, develop new products and services, and stay ahead of the competition.

Good measurement of the right data has some noticeable characteristics; the right data, collected correctly, is recognizable. It includes these characteristics:

■ Does the data tell us if our organizational strategy is working and well-aligned? Does it show us where to pivot or adjust to meet our goals and objectives?

- Does it energize us to action? (Do we get a "Eureka" moment – Ah-ha, I have an idea on how to fix or adjust this!)
- Does it allow us to better manage the process?
- Is it practical, easy to use, and not overly burdensome, expensive, or difficult to evaluate?
- Does it provide one obvious solution (A will improve or solve B) rather than multiple interpretations (We still don't understand what's causing B, so let's try C, D, and E…). Poor data quality results in inaccurate reflections of the current situation, leading to inappropriate decisions.

The right data implies that they must be relevant, accurate, timely, complete, and consistent. A brief definition of these data dimensions is provided below:

- *Relevancy*: Data relevance refers to the degree to which data is useful and applicable to a specific business need or problem. Relevant data is data that is directly related to the decision or question at hand and provides meaningful insights to inform decision-making.
- *Accuracy*: Data accuracy refers to how well the data reflects the true value or characteristic that it is intended to represent. In other words, accurate data is free from errors or biases and can be relied upon to make informed decisions or draw meaningful conclusions.
- *Timeliness*: Data timeliness refers to the degree to which data is available and up-to-date relative to the time of interest. In other words, it measures how current or recent the data is and whether it is relevant for the purpose it is being used. Timeliness is essential because decisions based on outdated or stale data may not reflect current realities and may lead to inappropriate or ineffective actions.
- *Completeness*: Data completeness refers to the degree to which all the necessary data elements are present and accounted for in a dataset. In other words, complete data includes all the information required for a particular analysis or task and does not contain any missing or erroneous values.
- *Consistency*: Data consistency refers to the degree to which data is uniform and conforms to established standards or rules. In other words, consistent data has a high degree of coherence, and the values are compatible. Data consistency can be assessed by checking whether the same data element is reported in the same way across various sources, time periods, or locations.

Establishing Measurement Objectives

To ensure you have a more focused plan to gather the "right" data, it helps to establish some measurement objectives so that you will not be wasting time gathering data that may not assist you in achieving your objectives.

The crucial thing you want to do is to determine the larger questions the data collection effort is intended to answer. If you are trying to understand why your team is not productive, some of the large questions may include:

- What are the team's goals and objectives, and are they clearly defined and communicated to all team members? (Per Chapter 2: Alignment)
- What are the roles and responsibilities of each team member, and are they clearly defined and understood by everyone? (Per Chapter 1: People and Culture)
- What is the team's work process, and are any inefficiencies or bottlenecks hindering productivity? (Per Chapter 7, Process Thinking)
- What is the team's communication and collaboration process, and is it working effectively? (Per Chapter 11: Human-Centered Organization – A Mindful Culture of Excellence)

Of course, this is not an exhaustive list! Any team might have different or additional parameters to consider.

The other important thing to define early on is how the data will be used for analysis. We have seen volumes of data with lots of analysis done. Still, the analysis came up needing to improve in terms of the insights that could be generated to answer the larger questions. By carefully defining how the data is used, you will narrow the scope of the data collection and, at the same time, prepare the analyst to develop an analysis plan. This will help define the analysis tools needed to prove or disprove your hypotheses.

The Mindfulness Reminder

Because thematically this book is concerned with the role that mindfulness plays in each of these organizational processes, let's touch on that now. Once again, the value of mindfulness may seem obvious, but as we've already said, common sense is not as common as we'd like to think, and likewise, sometimes mindfulness gets buried under routines and/or assumptions. Therefore, when looking for the "right" data to collect, we can see that relevancy, accuracy, timeliness, completeness, and consistency are all factors heavily influenced by the mindfulness of everyone involved.

Relevancy	Mindfulness helps us identify what is truly relevant by homing our focus and dismissing distractions
Accuracy	Mindfulness means better instructions are given, and fewer errors or misinterpretations are made
Timeliness	Mindfulness reduces delays
Completeness	Mindfulness helps us recognize whether our needed components are present or lacking
Consistency	Mindfulness, as always, keeps us aware and alert to our task, promoting our ability to be consistent in collection of data

Mindfulness plays an equally valuable role in establishing our measurement objectives, ensuring that everyone on the team understands goals, roles, the process, and communication between members.

Types of Measures to Consider

Chapter 5 discussed how a Suppliers-Inputs-Process-Outputs-Customers (SIPOC) could be used to identify the scope of the process we want to review and analyze. Depending on the scope of the analysis, the SIPOC can guide the collection of the data needed for the analysis.

Figure 8.3 shows a generic SIPOC (Suppliers, Inputs, Processes, Outputs, and Customers), which provides an overview of a high-level process, highlighting the key elements and interactions involved. This is an effective tool that can guide the collection of crucial data related to the process. It can also help identify important leading and lagging measures to monitor and analyze.

As depicted in Figure 8.3, the X's are your predictor (or leading) measures, and the Y's are your outcome (or lagging) measures. The Y's can be used to effectively define the major outcomes that will enable you to meet or exceed your customer needs, either internal or external, depending on the scope of your analysis. The X's reflect process activities or process inputs that enable you to achieve those process outcomes. The linkages among the inputs, processes, and outcomes are crucial to delivering what

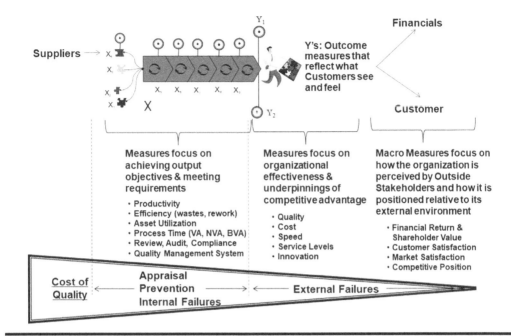

Figure 8.3   Data collection framework.

your customers expect. A careful review of these different measures, both lagging and leading, can be used to assess the various costs of quality, which include appraisal, prevention, and internal and external failure costs. A more detailed treatment on this subject can be found in *Principles of Quality Costs: Financial Measures for Strategic Implementation of Quality Management*, by Douglas C Wood.[1]

Data Collection Plan

The data collection plan, as provided in Figure 8.4, establishes objectives, specific output and input measures, measurement procedures, and a plan of action. This will help ensure that the "right" data are collected so that a more focused analysis can be performed to answer the broader questions about the problem being addressed.

An effective plan consists of the following elements:

■ **Measures and Operational Definition**: Critical output measures
   Y's and their related conditions X's are defined with clear operational
   definitions. To remove ambiguity and help all data collectors to have

Objective:		
Output Measures ("Y's" with operational definitions)	Related Conditions ("X's" with operational definitions)	Sampling Plan (if applicable)
Data Sources:		
Measurement Integrity: (Data Reproducibility & Repeatability)		
Action Plan: (who and when)		

Figure 8.4   Data collection plan template.

the same understanding and reduce the chances of disparate results between collectors, an operational definition must include:
- An exact description of how to derive a value for a measurable attribute that you are measuring
- A precise definition of the attribute and how, specifically, data collectors are to measure the attributes
■ **Data Source Considerations**: Data sources may be obtained from historical records or new data collection efforts need to be initiated. There are advantages and disadvantages to using historical data compared to the cost of collecting new data. Below are factors to weigh whether historical data should be used:
- Historical data may allow you to look at data covering a longer time span
- It may be more accessible or sometimes harder to obtain
- It may also be incomplete or unreliable
■ **Measurement Integrity**: Provide due diligence to ensure that the data collection process will result in the reproducibility and repeatability of the data being collected. This will help ensure accurate and reliable findings.
■ **Sampling**: Data collection is a time-consuming and costly endeavor. When necessary, it is important to exploit statistical sampling techniques to gather accurate and reliable data, while minimizing costs, time, and resources. Below are some tips to determine if sampling may be appropriate:
- The cost or time to gather 100% of the data is prohibitive
- It is too difficult to acquire 100% of the data

- The act of measurement destroys the object
- There is a need to cross-check historical data

By effectively using data and addressing the challenges associated with collecting and managing it, organizations can make more informed decisions, improve their operations, and gain a competitive edge.

What If There Is Insufficient or Little Data?

There are many situations when teams struggle to gather enough data for analysis. In a software development example, you may be dealing with the delivery of a software module every 6 months or so. Consequently, data on defects may be sparse. To deal with this problem, it helps to think about the bridge analogy, i.e., failure or collapse of a bridge will only happen every 100 years or so! Every bridge is somewhat different and takes a different amount of time to build. Empirically determining a bridge's reliability may take a hundred years (see Figure 8.5).

As depicted in Figure 8.5, the reliability of a bridge may take a very long time to determine (i.e., measuring the Y's). This is where it is helpful to think about the process of building a bridge and consider leading indicators (i.e., measuring the X's) that reflect the quality that goes into building a bridge:

■ The bridge reliability is a consequence of strong trusses and good welds
■ There are many opportunities to measure the strength of the welds and trusses – they are plentiful and repetitive

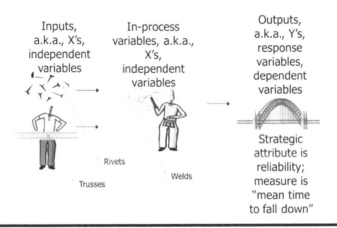

Figure 8.5   Bridge analogy for assessing quality when data is sparse.

Given the plentifulness and repetitive nature of these elements in the process, the gathering of measurement data on welds, rivets, and trusses to control the process is more readily available and you won't have to wait for the outcome (i.e., bridge failure) as it is too late to do anything about it.

Summary

Understanding data and its effective use can lead to better decision-making and improved organizational performance and enhanced customer experience. Effective collection and use of data reinforces a human-centered culture by showing consideration for those who use the data and make decisions from it. This can only be achieved when data are collected intentionally so that they can provide the foundation for evidence-based decision-making. The gathered data must be relevant, accurate, timely, complete, and consistent.

Questions for Discussion

1. What larger questions do you intend to answer with the data collection and analysis effort?
2. What is your plan to gather the "right" data, incorporating relevance, accuracy, timeliness, completeness, and consistency?
3. How are you using data when building and maintaining a human-centered culture?

Note

1. Wood, Douglas C., *Principles of Quality Costs: Financial Measures for Strategic Implementation of Quality Management*, 4th ed., ASQ Quality Press, Milwaukee, WI, 2013.

Chapter 9

Understand Variation

Variation is evil!

Jack Welch, Former CEO of GE

Imagine this: What would you feel if you put your right foot in an ice bucket and your left foot in some burning coal (see Figure 9.1)? Would you be feeling an average temperature of freezing ice and burning coals? Or would you feel the temperature variation between these extremes (e.g., freezing ice and burning coals)? Of course, you would feel the variation! And that is precisely what your customer would feel! And, if this is how your customer feels day in and day out from using your product or service, how long would it take before they stop buying from you and move to an alternate vendor? I would not even give it a second chance!

Hung Le: *In a real case scenario a few years back, I presented to an executive team at a 300-bed hospital. We reviewed their patient satisfaction score. Right when I showed them the score chart, one of the administrators jumped up with excitement and said, "That's wonderful!" The patient satisfaction score rose from 85% in the previous quarter to 92% in the current quarter.*

I asked: "What has the organization done to drive this increase?"

He replied, "Just normal execution. Nothing special."

Sure enough, their patient satisfaction score dropped to 87% the following quarter.

The observed variation was random and was inherently a part of normal fluctuations in the process!

DOI: 10.4324/9781003454892-9

Figure 9.1   Does it all average out?

The example above shows how organizations tend to react to a change between two numbers. For one obvious reason, a change in either direction between two data points does not establish a good or bad trend. Second, it may not be the result of an intentional act the organization is doing and, therefore, is part of the normal variation expected in any process. With a proper understanding of the drivers of variation in a process, quality will improve, and quality improvement can be accomplished. As W. Edwards Deming put it: "If I had to reduce my message for management to just a few words, I'd say it all had to do with reducing variations." Process variations are the primary cause of poor quality.

In this chapter, we define process variation and why it is crucial to understand it. Then, we introduce a control charting tool to differentiate the two types of variation. This is used to assess your process stability and capability and determine an appropriate course of action to improve the process.

What Is Process Variation?

When the steps of a process vary, the output of the process varies. The more a process varies, the more the output of that process varies. This is a leading cause of quality issues because of the high variability observed in the output of the process, which renders it unpredictable. Figure 9.2 shows examples of how spread out the data are depending on the level of variability observed.

Figure 9.2   Examples of data distribution.

Everyone should first realize that all products or processes are unique[1]. Think about when you need to drive from point A to point B. It will normally take you, on average, 30 minutes. But the time you arrive at your destination will generally vary by a few minutes from day to day. Depending on the different factors, the amount of traffic you encounter, the weather, the red lights versus green lights, or getting behind the neighborhood school bus (or avoiding it) – any of these factors and more can add or subtract minutes. This is what is meant by process variation. After enough experience, you can probably predict your driving time to within ±10 minutes. Or even within ±5 minutes.

Of course, ideally, you would want no variation at all, and you could predict with 100% certainty that your driving time is 30 minutes every time! But that's not how commuting works, and it's not how most processes work, either. There will always be some amount of variation. When we can predict variation, we can plan for it ahead of time or cope with it better when it does happen. When we understand what variables are at play, we can reduce or eliminate *some* of them.

Any given process contains many sources of variation. While differences may be large or small, variation is always present. It is a naturally occurring phenomenon inherent within any process. The quality of any products or services in question here is whether the variation inherent in a product or service is within the customers' acceptance level.

Process variation is a measure of dispersion, or how spread out the data are. And it is often expressed as the data's range, variance, or standard deviation. Think back to your classes in high school or college. Remember what the class average score was for your class, how the score spread around this average, and how well you did against the class average. If your score was two standard deviations above the average, then congratulations! You must have received an A (or A+, depending on how much of an "outlier" you were!). And if you were one standard deviation below the average, you had some catching-up work!

Your instructor's goal would have been for the class to attain a high average with a tight spread around the average. Assuming that course content and tests were designed well, this would demonstrate a high level of understanding by the entire class. If this were not achieved, the class instructor would have to determine the factors that drove a low average and a large spread for the class and improve the overall class performance.

Why Is Variation Important?

Hung Le: *When I was working for GE back in the mid-'90s, Jack Welch was at the helm. I heard many of his speeches. One phrase that I remembered most was: "Variation is evil!" It was something that Jack was deeply passionate about driving variation out of GE processes because that was how GE could excel at giving their customers what they wanted the first time, every time!*

As discussed earlier, quality management is a lot about how predictable your process is. Simply put, the aim is to perform processes in such a way that the outputs are repeatable with consistent results. In doing so, we can better manage, control, and improve them, as required, to meet or exceed our customers' expectations.

When a process experiences a high amount of variability, quality issues arise, and root problems must be determined. This challenge is being fundamentally addressed by the first rule implemented in the Toyota Production System (TPS), which states that all work shall be highly specified according to the content, sequence, timing, and outcome. For example, when a car's seat is installed, the bolts are always tightened in the same order, the time it takes to turn each bolt is specified, and so is the torque to which the bolt should be tightened. These work processes consist of many steps carried out by people who have specific positions in a team structure. Hence, workflows should be clearly defined so that each person knows exactly what they should be doing and when they should do it. Process variations are reduced when roles and responsibilities are clearly defined and implemented.

This TPS rule removes the most significant source of variation in the process, so any deviation in timing and the expected outcome can be analyzed, root causes can be more readily isolated and determined, and the implemented process can be improved. Figure 9.3 shows how process performance can be improved by decreasing process variation. And less

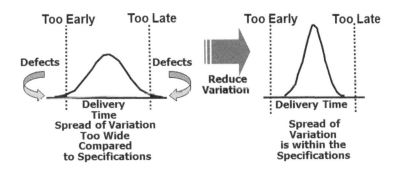

Figure 9.3   Benefits of reduced process variation.

variation results in greater predictability of your process; less waste and rework, which lowers costs; better and long-lasting products and services; and happier customers.

Two Types of Variation

There are two types of Variation: Common Cause and Special Cause. Understanding and differentiating these variations is essential because a different strategy will be required to reduce and eliminate every kind of variation.

- ■ Common Cause Variation
  When you hear the term "common," it implies that something occurs fairly frequently, and there is a high chance that it will happen again in the near future. Common Cause Variation is caused by random fluctuations in the system resulting from many small contributing factors. So, you can predict it within a specific range and have a high chance of being correct. Common Cause Variation shares the following characteristics:
  - – The variation that comes from within the process
  - – Predictable within a range
  - – The remedy is to change how work is performed
  The last trait has to do with the fact that no single factor contributes to the process variation due to its random nature. The only way to address and reduce it is by focusing on how the work is performed and improving it.
  Example: Variation in work commute due to traffic lights, pedestrian traffic, and parking issues.

■ Special Cause Variation

Special cause variation consists of a "special" event that is rare and likely unique or one-of-a-kind. Consequently, it cannot be predicted or can be predicted at extremely low probability. Therefore, Special Cause Variation is caused by large fluctuations in the system which result from a single factor (i.e., an assignable cause). Special Cause Variation has the following characteristics:
  – The variation that comes from outside the process
  – Unpredictable
  – The remedy is to insulate the process

Because a single factor causes Special Cause Variation, the remedy to eliminate it is to isolate this single root cause and implement some countermeasures to prevent it from recurring.

Example: Variation in work commute impacted by a flat tire, road closure, or heavy frost/ice.

Consider this question: Is the sale revenue for December at a retail store a common cause or special cause variation? Why?

What Is a Control Chart?

One of the tools we often use to manage process variation is a control chart. Simply put, it is:

■ A simple graphical representation of the performance of a process over time
■ A set of horizontal lines, known as control limits, are calculated based on the process variation, which delineates an upper and lower limit of the acceptable range of the results of a process
■ Used to detect "unusual" signals, i.e., something that is not random

The control limits are set, usually ±3 standard deviation by default. This implies that there is still about a 1% chance that a data point will exceed the control limits and not be a special cause (i.e., 1% false alarm – determine a special cause when it's not) as one lowers the control limits, the percentage of false alarms increases. And, as one increases the control limits, the percentage of mis-detecting a special cause increases when it exists (see Figure 9.4).

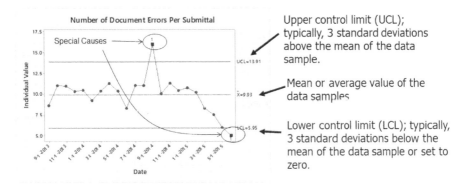

The following labels appear to the right of the chart:

Upper control limit (UCL); typically, 3 standard deviations above the mean of the data sample.

Mean or average value of the data samples

Lower control limit (LCL); typically, 3 standard deviations below the mean of the data sample or set to zero.

Figure 9.4   Control chart with special causes.

Consider this scenario: Is a high false alarm rate at a security screening checkpoint in an airport (i.e., deciding that someone is carrying a weapon when they are not) more severe than *failing* to detect someone who is actually carrying a weapon?

What Are Some Signals of a Special Cause Variation?

Figure 9.5 shows different types of signals of special cause as a sign of process instability.

Any time when there exists a special cause variation, it indicates that there is something unusual going on and the process needs to be investigated further to identify the cause of the deviation and take corrective action if necessary.

In Figure 9.5:

■ When Test 1 is violated, it usually is a sign something unusual happened in the process and it needs to be investigated.

Test	Description
1	1 or more points beyond the upper or lower control limits
2	9 consecutive points on one side of the center line
3	6 consecutive points either increasing or decreasing
4	14 consecutive points alternating up and down

These are the common signals; others exist

Figure 9.5   Different types of special cause signals.

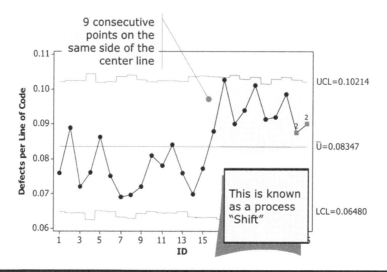

Figure 9.6 Control chart showing a process shift.

■ When Test 2 is violated, it implies that a process shift has occurred. This could be good or bad. Please refer to Figure 9.6 and discussion of process shifts below.
■ When Test 3 is violated, it implies that a trend, increasing or decreasing, exists. Please refer to Figure 9.7 and discussion of special trends below.
■ When Test 4 is violated, it normally reflects an over control pattern. This shows that the process is affected by some external factors that operate on a regular basis, such as alternating raw materials, a seasonal effect, a daily or weekly routing, or a shift in personnel.

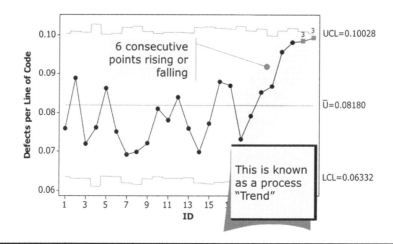

Figure 9.7 Control chart showing a special trend.

As depicted in Figure 9.6, a process shift occurred. Whether the shift is good or bad depends on what has been done in the process. Below is an explanation of the meaning of a shift:

- ■ Shifts are good where …
  - – Response to a deliberate process change, and
  - – Run is on the desired side of the centerline,
  - – This is compelling evidence to claim improvement victory.
- ■ Shifts are bad where …
  - – You have done nothing to cause a change in performance.
  - – The run is on the unwanted side of the centerline.

Figure 9.7 exhibits a special trend. An explanation of whether a trend is good or bad is provided below:

- ■ Trends are good where …
  - – Direction is favorable.
  - – The response is to a deliberate process change.
  - – There is compelling evidence that the process is seeking a new and better level of performance.
- ■ Trends are bad where …
  - – Direction is unfavorable.
  - – You need to find out why it is occurring.

What Does All This Mean about Your Process Status?

You often hear patient care practitioners, either doctors or nurses, talk about a patient having stable vital signs. In general, this means the person's vital signs – like their heart rate, blood pressure, and body temperature – are steady and within normal limits. They're conscious (aware) and comfortable. In the context of process performance, a process is stable when it is steady and is within "normal" limits. This implies that it only exhibits common cause variation, as opposed to an unstable process, which shows signs of special cause variation. Process capability, on the other hand, has to do with whether a process can meet and exceed the customer's requirements (see Figure 9.8).

Another example is when a police officer stops you when he sees your car swerving, perhaps due to DUI. He then asks you to breathe into a breathalyzer and walk in a straight line (or as straightforward as possible). Of course, you will attempt to stay in a straight line as much as possible. If

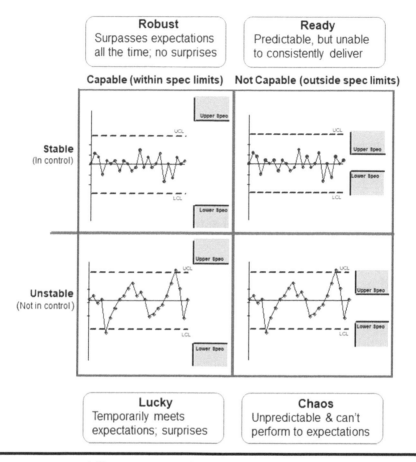

Figure 9.8 Process stability versus capability.

you veer too far left or too far right or fall, you will be charged with DUI because you will have little control of your vehicle and can cause accidents!

- What you would desire is for a process to be robust.
  - Robustness is the intersection of stability and capability.
  - Robust processes reliably produce what they are intended to produce.
  - Robust processes do not just happen. They are engineered.

This is where you would want your process to be – stable and capable. In other words, it can reliably produce what it is intended to create. The customers will receive their product/service on time every time! That is how you can build customer trust and loyalty.

When dealing with an unstable process, the first order is to stabilize it (i.e., remove any special causes) regardless of whether it is capable. You must do so to improve it and make it predictable. Once a process is stable, it is time to make it capable by reducing variation in your process (i.e., remove common causes and tighten the process). In other words, as shown in Figure 9.8, you would want to move from the Chaos state to Ready state first. Then, capability can be achieved from the Ready state to the Robust state.

Tips for monitoring and managing your process:

■ Process not stable – showing signs of special cause.
   – Use the tools of problem solving to identify the root cause.
   – If a special cause impacts the process negatively, develop procedures to eliminate the return of the problem. If beneficial, implement procedures to make them permanent.
■ Process stable – showing only common cause variation.
   – Use the tools of improvement to study all the data and identify factors that cause variation.
   – Determine what needs to be changed to improve performance and predictability.

Questions for Discussion

1. Discuss the impact of variability in products and services on how employees view the performance of their organization. Should employees care about the quality of what they produce?
2. Is your process stable? Are there any special causes or trends that render the process unstable? If so, what is driving those special causes? What countermeasures can be implemented to stabilize the process?
3. If the process is stable, how capable is the process? If the process cannot meet the customer's requirements, what factors are driving the process variation? Determine what needs to change to improve the process capability.

Note

1. Wheeler, Donald J., *Understanding Variation: The Key to Managing Chaos*, SPC Press Publisher, 2nd edition, 2000.

Chapter 10

The Basic Quality Toolbox

If ALL you do have is a hammer, you better use it.

TJ McCue, Senior Contributor, Forbes.com

Grace Duffy: *As a degreed Archaeologist, I know that the Egyptians used quality tools to build the pyramids. The compass and the square are basic symbols in ancient history and are still used in some engineering graphics (see Figure 10.1 for an image of tool use in ancient Egypt). The evolution of quality tools parallels the evolution of quality itself. Tools are simply vehicles for getting something done "easier, better, faster, cheaper." This chapter drills down to the basic quality tools, but keep in mind that the best tools are no good without the best people (Chapter 1) and the best processes (Chapter 7).*

The history of quality reaches back into antiquity. The current quality movement began in the 1920s. The "quality profession," as it is now known, started with Walter Shewhart of Bell Laboratories. He developed a system known as statistical process control (SPC) for measuring variance in production systems. SPC is still used to help monitor consistency and diagnose problems in work processes. Shewhart also created the Plan-Do-Check-Act (PDCA) cycle, which is a systematic approach to improving work processes. When the PDCA cycle is applied consistently, it can result in continuous process improvement.

Kaoru Ishikawa first pulled together the "Basic Quality Tools" in his book *What is Total Quality Control?*[1] Walter Shewhart and W. Edwards Deming began developing the initial quality improvement tools in the 1930s and 1940s. This development resulted in a better understanding of processes and led to the expansion of the use of these tools in the 1950s.

                                            DOI: 10.4324/9781003454892-10

Figure 10.1 Compass and the square used as tools in building of the pyramids.

The Japanese began to learn and apply the statistical quality control tools and thinking taught by Ishikawa, then head of the Union of Japanese Scientists and Engineers (JUSE). These tools were further expanded by the Japanese in the 1960s with the introduction of the following seven classic quality control tools:

- Cause-and-effect diagram
- Check sheet
- Control chart
- Flowchart
- Histogram
- Pareto chart
- Scatter diagram

As economies and organizations matured, quality frameworks, such as the Malcolm Baldrige National Performance Excellence program, ISO, Lean, and Six Sigma, were developed to encourage systematic approaches to integrating quality practices into all areas of the business. More recently, the use of Industry 4.0 has become an enabler for human/process effectiveness. This re-humanizing of the process-centric approach of the 20th century is a welcome enhancement for the 21st century. Aligning tasks to the priority outputs and outcomes of the organization sets an invigorating direction for employees and leaders.

Tools are a means to an end. Developing and sustaining a human-centered organizational culture should focus on the following sequence "Simplicity, then stability, then quality, then capacity, then velocity, then cost." This sequence is critical in using tools that reinforce human decision-making based on VoC and continuous feedback, as well as accurate data. Tools enhance the ability of engaged, enthused individuals to meet and exceed the VoC. They help us sustain the cycle of observe, orient, design, analyze, and improve.

Quality specialists tend to gravitate to sets of tools that are most comfortable for their working styles. Our general guidance is to use a sequence of brainstorming ideas or problem symptoms, using an affinity diagram to organize the ideas into themes, then using the themes as major categories in the cause-and-effect diagram, also known as the fishbone/ Ishikawa diagram. This set of three tools is excellent for guiding the exploration of symptoms of a problem or opportunity.

Many of the quality tools do not give answers, they simply provide the foundation for informed discussion around an issue. A good quality facilitator will use these tools to engage a team in problem-solving and decision-making based on established goals. Aligning team discussion to priority goals connects individuals to the organization. Seeing results that directly support their jobs is a strong energizer for most employees. As discussed in earlier chapters, engaged, enthusiastic employees come together to sustain a human-centered organizational culture.

Another set of three tools useful in your improvement projects is described in the following examples:

- Pareto analysis
- Fishbone/Ishikawa diagram
- Failure Mode and Effects Analysis

Pareto Analysis

The Pareto principle or 80/20 rule is adapted from earlier economic works of Vilfredo Pareto (see Figure 10.2; 20% of the population had 80% of the wealth). Dr. Joseph Juran is credited with connecting the concept to data analysis and

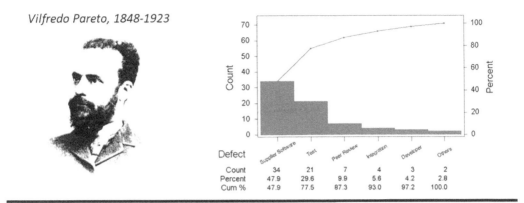

Vilfredo Pareto, 1848-1923

Defect	Supplier Software	Test	Peer Review	Integration	Developer	Others
Count	34	21	7	4	3	2
Percent	47.9	29.6	9.9	5.6	4.2	2.8
Cum %	47.9	77.5	87.3	93.0	97.2	100.0

Figure 10.2   Vilfredo Pareto and Pareto chart.

process improvement. As applied to process improvement, the Pareto principle states that only a few vital factors are responsible for producing most problems. The key to success is to concentrate on correcting those factors first.

For example:

■ 80% of traffic jams occur on 20% of the roads. Focus solutions on those roads.
■ 80% of beer is consumed by 20% of drinkers. Advertise heavily to those customers.
■ 80% of profits come from 20% of customers. Follow up diligently with those customers.
■ 80% of sales are generated by 20% of salespeople. Reward those salespeople.

The focus question is: Where should we spend our time and resources on improvement? A Pareto chart is a visualization tool that draws the eye to the most frequently occurring defect, observation, cost, etc. Pareto charts help you narrow down your focus to get to the one thing that needs to be fixed; it helps cure absence blindness.

Pareto Chart Features

Figure 10.3 is an annotated example of a Pareto chart showing errors identified during the software development process. In this case, close to 50% of the errors discovered are attributed to supplier software. The chart identifies the highest priority area to address for error resolution. A caution in the interpretation of a Pareto chart is that the frequency of error is not always the only gauge. Depending on the cost of the error to the company

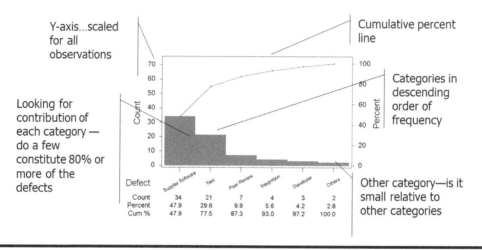

Figure 10.3   Pareto chart features, demonstrated by software errors and the point at which the errors are found.

or its impact on the customer, a less frequent, but much more expensive defect may be the best to tackle first. This highlights the importance of a human-centered culture. We don't just blindly follow the tool or technology. We align the data and analysis to the strategic objectives of the organization. Interpretation of data must be influenced by the priorities of the organization based on its mission and VoC.

Figure 10.4 reinforces the need for a human-centered interpretation of the Pareto chart. What if you see counts of almost equal frequency in the chart? It may be that a different value other than frequency, such as cost,

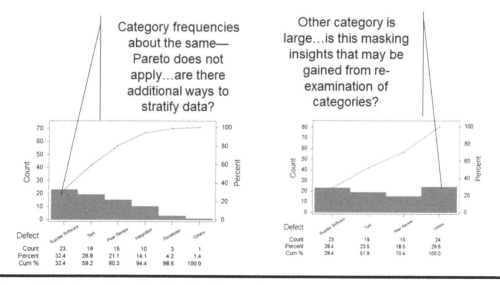

Figure 10.4   Examples of ineffective Pareto charts.

or customer impact will provide a better analysis. Or if the "other" category in the Pareto is larger than any of the titled categories, it will be necessary to re-cut the data to provide more granularity. Sometimes you must apply Pareto analysis more than once to home in on the area of interest.

Cause-and-Effect Analysis

The cause-and-effect diagram graphically illustrates the relationship between a given outcome and all the factors that influence the outcome. It is sometimes called an Ishikawa diagram (after its creator, Kaoru Ishikawa) or the fishbone diagram (due to its shape).

The cause-and-effect diagram (see Figure 10.5) displays the factors that are thought to affect a particular output or outcome in a system. The factors are often shown as a grouping of related sub-factors that act in concert to form the overall effect of the group. The diagram helps show the relationship of the parts (and subparts) to the whole by:

- Determining the factors that cause a positive or negative outcome (or effect).
- Focusing on a specific issue without resorting to complaints and irrelevant discussion.
- Determining the root causes of a given effect.
- Identifying areas where there is a lack of data.[2]

A related tool to the cause-and-effect diagram is the "5 Whys." Once groupings are displayed on the "fishbone" structure of the diagram, the

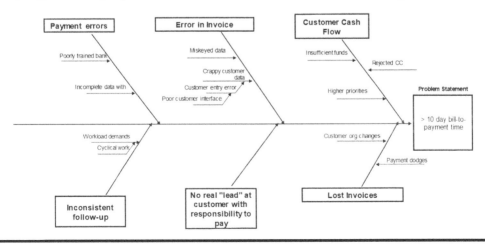

Figure 10.5   Fishbone, cause-and-effect, or Ishikawa diagram example.

highest priority defect or observation is addressed by asking a series of Why questions to drill down to an actionable cause that can be resolved. An example of the 5 Why tool follows:

The National Park Service was concerned that frequent washings of the Jefferson Memorial were causing the marble used at the memorial to degrade over time. The 5 Whys were used to determine the real root cause.

- Why were frequent washings required? Because there were bird droppings.
- Why were there bird droppings? Because the birds feed on the spiders at the memorial.
- Why were there spiders at the memorial? Because there were midges (insects) at the memorial.
- Why were there midges at the memorial? Because they were attracted by the memorial lights at dusk.

The solution: Keep the lights turned off until an hour after dusk. No lights at dusk meant fewer midges. Fewer midges reduced the number of spiders. Fewer spiders meant fewer birds and a reduced amount of bird droppings and less frequent washings. This kind of repetitive thinking can help you focus on the real root cause(s) of your problem.

General steps for creating a cause-and-effect diagram follow.

Cause-and-Effect Diagram Checklist

1. Review the focused problem
2. Identify possible causes through brainstorming/brainwriting
3. Sort possible causes into reasonable groupings via an affinity diagram
4. Choose a grouping and label a main "fishbone"
5. Develop and arrange fishbones for that grouping
6. Develop other main fishbones
7. Add title, date, and contact person
8. Select possible causes to verify with data

Fishbone/Ishikawa Diagram Example

Process:

1. Start with a focused problem statement
2. Use brainstorming or brainwriting to develop ideas that provide "causes"

3. Group by relationship – the affinity diagram is a useful tool here
4. Go as deep as required to focus on the root cause(s)

Remember to use causal thinking to develop the fishbones.
Example: Bill-to-Pay Cycle Time

Failure Mode Effect Analysis (FMEA)

Failure Modes and Effects Analysis (FMEA) is a systematic tool used for analyzing systems, designs, or processes to identify potential failure and its causes. It aims to identify and reduce the risk of failure by focusing actions on areas of greatest risk. There are two types in general use: design FMEA (DFMEA) for analyzing potential design failures, and process FMEA (PFMEA) for analyzing potential process failures. FMEA is for *potential* failures and is less useful for current problems. Should a situation be occurring, the root cause analysis tool is preferred.[3]

The FMEA tool can be used in several situations:

∎ When changing a product, process, service, or policy (Improve/Control in DMAIC or Analyze/Design in DFSS).
∎ When trying to narrow the project scope (Measure in DMAIC) and when identifying "soft" x's (Analyze in DMAIC).

For …

∎ Identifying specific ways in which a product or process step may fail.
∎ Develop countermeasures targeted at those specific failures that will improve the performance, quality, reliability, or safety of the solution.

How to create an FMEA?

1. Identify potential failure modes and ways in which the product, service, or process might fail.
2. Identify the potential effect of each failure (consequences of that failure) and rate its severity.
3. Identify the causes of the effects and rate their likelihood of occurrence.
4. Rate your ability to detect each failure mode.

5. Use the Action Priority (AP) Matrix to determine the risk of each failure mode.
6. Identify ways to reduce or eliminate risks associated with high APs.

Note that in 2018 the Automotive Industry Action Group (AIAG) developed a new process for creating an FMEA using the Action Priority Matrix (see Figure 10.7) rather than the traditional Risk Priority Number (RPN) that multiplied the severity, occurrence, and detectability (S.O.D.) ratings. The Action Priority Matrix considers the S.O.D. independently, on a comparative scale. RPN multiplication dilutes individual ratings and creates a single number. The concern which prompted the AIAG update was that a severity of 10 might be overshadowed by a low frequency or detection score.

The SIPOC-Flowchart-FMEA Sequence for Failure Mode and Effects Analysis

Figure 10.6 illustrates three tools useful in sequence for setting up an FMEA. The SIPOC is a macro-level analysis of the suppliers, inputs, processes, outputs, and customers of a process. The SIPOC discussion identifies the components of a process, its inputs, who supply them, the major process steps, the outputs, and who uses those outputs. Each of these components provides opportunities for errors or defects to enter the process. The SIPOC is a high-level flowchart that is then expanded into a full process map. The team documents potential areas of failure as they identify steps and resources involved in each component of the process map.

Figure 10.6 continues by illustrating the VoC Key Issues and Critical to Quality (CTQ) items identified by the customer, along with their specifications. This information further suggests priority potential effects of failure to be anticipated for recommended action.

Figure 10.7 illustrates a full FMEA Template. FMEA is carried out by a team whose purpose is to look for all ways a product or process can fail, also called "failure modes." Each failure mode may have one or more effects, some more likely to occur than others. Each effect has its own risk and potential for severity. The FMEA process is a method of identifying potential failures, the effects of those failures, and the associated risks of a product or process to eliminate or reduce them.

You will see among the final columns on the right that potential errors must be rated in three dimensions: severity, occurrence (or frequency), and

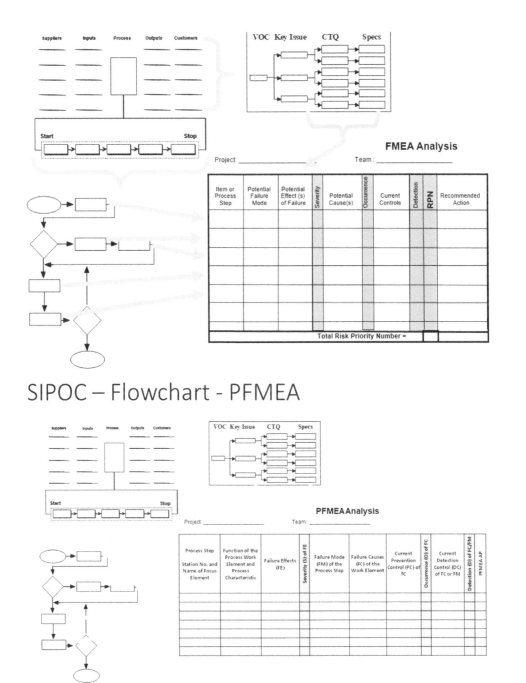

SIPOC – Flowchart - PFMEA

Figure 10.6   Process tools for identifying potential failures and effects.

detectability. These three scores are scaled from one to ten, as suggested in Figures 10.7a–c, then compared with an Action Priority Matrix to establish a high, medium, or low risk. The higher the risk assessment, the more pressing it is to find a solution!

1. Process Item	2. Process Step	3. Process Work Element	1. Function of the Process Item	2. Function of the Process Step and Product Characteristic	3. Function of the Process Work Element and Process Characteristic	1. Failure Effects (FE)	Severity (S) of FE	2. Failure Mode (FM) of the Process Step	3. Failure Causes (FC) of the Work Element	Current Prevention Control (PC) of FC	Occurrence (O) of FC	Current Detection Control (DC) of FC or FM	Detection (D) of FC/FM	PFMEA AP	Special Characteristics	Filter Code (Optional)	Preventive Action
System, Subsystem, Part Element or Name of Process	Station No. and Name of Focus Element	4M Type	Function of System, Subsystem, Part Element or Process	(Quantitative value is optional)													

Figure 10.7   FMEA template.[4]

Severity is a rating that corresponds to the seriousness of an effect of a potential error mode. An error that can endanger human life, health, or safety would certainly warrant a rating of 10 or at least 9 – but of course every process has its own specific traits. Typographical errors in a document might be considered a "very slight" hazard for a romance novel, a "moderate" hazard for a book of recipes, but a "serious" hazard for a drug usage guide or a safety manual.

Occurrence or Frequency is a rating that corresponds to the rate at which a first-level cause and its resultant error mode will occur over the cycle of a process. Remember that this is a rating of the likelihood that an error will occur at the point of its origin – it is *not* the stage at which these errors might be noticed. If we notice that 5% of our widgets are too big to

Value	Description
10	Hazardous
9	Serious
8	Extreme
7	High
6	Significant
5	Moderate
4	Minor
3	Slight
2	Very Slight
1	None

Severity is a rating that corresponds to the seriousness of an effect of a potential error mode

Figure 10.7a   Severity scale.

Value	Description
10	Almost Always
9	Very High
8	High
7	Moderate High
6	Medium
5	Moderate
4	Slight
3	Very Slight
2	Remote
1	Almost Never

Figure 10.7b   Occurrence or frequency scale.

Value	Description
10	Almost Impossible
9	Remote
8	Very Slight
7	Slight
6	Low
5	Medium
4	Moderate High
3	High
2	Very High
1	Almost Certain

Figure 10.7c   Detectability scale.

fit in their pre-manufactured boxes, that is a matter of detection. We must travel backward in the process to find the root cause, or the point at which the error that causes a size discrepancy is introduced into the system, and the likelihood that this root cause will happen.

By the way, if you read this example and thought, "Maybe the pre-manufactured boxes are the source of the problem," then congratulations on your own mindfulness! It's not necessarily the correct answer to this imaginary scenario, but if we're looking for the root cause, it's an option worth considering.

Detectability is a rating that corresponds to the likelihood that the detection methods or current controls will detect the potential error mode before the outputs of the process are released to the customer.

Detectability is an interesting trait to consider. If we have an error with a very low detectability, we might consider finding ways to increase our detection, especially if the resulting severity is high. Leaky gas pipes should definitely be detected before they are installed, or lives could be lost. Does there need to be more inspection? Additional tests or comparisons?

On the other hand, we know that adding "detection" measures means adding another step (or steps) into a process, while our overall goal is to simplify. Does the error have a root cause we can adjust or eliminate? Are there some errors that are simply unavoidable? Each process will have its own parameters that answer these questions.

Action Priority (AP) tables emphasize Severity, Occurrence, and Detection (S.O.D.) in that order, to emphasize prevention controls. Figure 10.7d is an example of an AP table.

The AIAG/VDA FMEA manual uses SOD tables found in the "Blue & Red Book". There is still a substantial RPN legacy in industry as teams transition to AP formats. Note the SOD levels for AP levels of Medium and Low.

S	O	D	AP	Justification for Action Priority - PFMEA
9-10	6-10	2-10	H	High priority due to safety and/or regulartory effects that have a high or very high occurrence rating
9-10	4-5	7-10	H	High priority due to safety and/or regulartory effects that have a high or very high detection rating
5-8	4-5	5-6	H	High priority due to the loss or degredation of a primary or secondary vehicle function or a manufacturing disruption that has a moderate occurrence rating and a moderate detection rating
5-8	4-5	2-4	M	Medium priority due to the loss or degredation of a primary or secondary vehicle function or a manufacturing disruption that has a moderate occurrence rating anda low detection rating
2-4	4-5	2-4	L	Low priority due to perceived quality (appearance, sound, haptics) or a manufacturing disruption with a moderate occurance and low occurance rating

Figure 10.7d   Action priority matrix translation.

Figure 10.8 is a completed process FMEA for ATM errors. The goal is to identify the highest priority or impact potential failures by analyzing the severity, occurrence, and detectability values to get an Action Priority (AP) value. The highest value is the first failure to be addressed. In the case of Figure 10.8, the most pressing problem is money not being disbursed because the machine runs out of cash. The improvement team recommended an action to increase the amount of cash stored in heavily used ATMs to prevent out-of-cash instances. In many FMEA templates, there are an additional four columns added to the right of the Excel spreadsheet to allow for a second rating of severity, occurrence, and detectability, or AP, after the corrective action has been taken. The goal is for the second assessment of AP to be greatly lowered or even eliminated due to the action taken. You can see that the FMEA tool is a communication and discovery device that requires human-centered involvement by engaged subject matter experts to be effective. Unengaged, unmotivated individuals will not put the level of effort into the discovery or corrective action required to gain a successful resolution.

Input from the team is definitely required in determining "cutoffs" for which issues must be addressed. If a Severity is extreme, the error must be hastily addressed. For example, based on their knowledge of the process, teams can determine at what point a severity score must be addressed, regardless of the other two scores. The same principle applies for the other factors as well – an error with an extremely high occurrence, despite low severity and low detectability, seems like an obvious opportunity for improvement.

1. Process Item (System, Subsystem, Part Element or Name of Process)	2. Process Step (Station No. and Name of Focus Element)	3. Process Work Element (4M Type)	1. Function of the Process Item (Function of System, Subsystem, Part Element or Process)	2. Function of the Process Step and Product Characteristic	3. Function of the Process Work Element and Process Characteristic (Quantitative value is optional)	1. Failure Effects (FE)	Severity (S) of FE	2. Failure Mode (FM) of the Process Step	3. Failure Causes (FC) of the Work Element	Current Prevention Control (PC) of FC	Occurrence (O) of FC	Current Detection Control (DC) of FC or FM	Detection (D) of FC/FM	PFMEA AP	Special Characteristics	Filter Code (Optional)	Preventive Action
ATM System	PIN Authentication	Personnel	User identification	Monitor for unauthorized access	Prevent unauthorized access	Very dissatisfied customer	8	Withdrawl not authorized	Lost or stolen ATM card	Block ATM card after three failed authentication attempts	3	Comparison of ATM card number to known lost or stolen cards	3	L			
ATM System	Dispense cash	Machine	Dispense cash	Open ATM slot and move cash to external dispenser	Get approved cash to dispenser user	Dissatisfied customer	7	Cash not dispursed	ATM out of cash	Internal alert of low cash in ATM	7	Assigned individual to monitor alerts	4	H			Increase minimum cash threshold limit of heavily used ATMs to prevent out-of-cash instances

Note: Validation columns to the right of Preventive Action truncated for display purposes

Figure 10.8   Example of a completed process FMEA for automated teller machine errors.

Summary

This chapter describes the basic quality tools and shares their value in organizational communication and discovery. Quality tools enhance the ability of individuals to gather data, analyze situations, brainstorm potential solutions, and test options for resolution. A human-centered organization values the input of its employees as subject matter experts. Individuals are motivated within a culture that respects their involvement and treats them as significant contributors to company outcomes. The three tools described as examples in this chapter are offered for your use in improvement projects:

- ■ Pareto analysis
- ■ Fishbone/Ishikawa diagram
- ■ FMEA

Questions for Discussion

1. Discuss the evolution of quality tools from antiquity to now. How have the tools changed, and how might they have stayed the same?
2. Identify how quality tools can be used to generate effective communication among stakeholders and team members in support of continuous improvement and problem solving.
3. Explore how potential failures can be identified as a team builds a flowchart or process map. What happens in our minds as we envision steps necessary to perform a task or process?

Notes

1. Ishikawa, Kaoru, *What Is Total Quality Control? The Japanese Way,* Prentice Hall, Inc. Englewood Cliffs, NJ, 1985.
2. Duffy, Grace L., editor, *The ASQ Quality Improvement Pocket Guide*, ASQ Quality Press, Milwaukee, WI, 2013, p. 50.
3. Duffy, Grace L., and Furterer, Sandra L. *The ASQ Certified Quality Improvement Associate Handbook*, 4th ed., ASQ Quality Press, Milwaukee, WI, 2020, pp. 201–202.

4. Template taken from QIMacros application available at QIMacros.com. The template is consistent with the Automotive Industry Action Group. The AIAG & VDA FMEA Handbook is the new automotive industry reference manual for Failure Mode and Effects Analysis; it is to be used as a guide to assist suppliers in the development of Design FMEA, Process FMEA, and Supplemental FMEA for Monitoring and System Response. Developed with a global team of OEM and Tier 1 Supplier Matter Experts (SMEs), it incorporates best practices from both AIAG and VDA methodologies into a harmonized, structured approach.

Chapter 11

Human-Centered Organization: A Mindful Culture of Excellence

You cannot manage other people unless you manage yourself first.

Peter Druker

In this chapter, we discuss the subject of mindfulness. This valuable human skill – of full and non-judgmental awareness of one's present state – is proving its benefits in countless areas in our lives. We are certain this is not the first time you've heard the word or the philosophy! And obviously, we cannot encompass the whole topic of mindfulness in one chapter of one book. Mindfulness is studied and practiced worldwide and the information about its benefits is voluminous.

However, we do want to discuss how mindfulness as a practice can benefit organizations and the people within them. Figure 11.1 depicts the contrasting states of an individual engulfed by their thoughts, memories, and concerns about the future, as opposed to a state of mindful awareness and presence in the present moment.

Have you ever …

- Said something you later regretted?
- Made a decision that went against your values?
- Upset a coworker or someone you cared about and were unaware of your action?

DOI: 10.4324/9781003454892-11

Figure 11.1 Is your mind full, or are you mindful?

- Been working hard and did not feel like you are making progress?
- Got frustrated because someone talked to you disrespectfully?
- Lost sleep because you were afraid you might not be able to meet a project milestone?
- Walked into a room to get something and then needed to remember what you were looking for?
- Have you read something and, after a couple of pages, had no idea what you've read?

These are just a few examples of how we let our minds run on autopilot, unaware of what is happening to each of us and our surroundings! Without self-awareness and emotional regulation skills, we face various personal and professional obstacles.

These issues are writ large when an organization doesn't practice mindfulness. An organization functions as an "individual" entity that can suffer from the same ramifications: ignorance of the consequences of actions, failure to make progress (spinning the wheels), frustration and disrespectful behavior; barely (or not) meeting requirements, forgetfulness, wastefulness, and the general disinterest of all the workers involved. Everybody is so preoccupied and overloaded with information that no one can determine what is important, what could be improved, and what valuable resources are going to waste.

Some of the examples may include:

1. Your leadership and staff may be challenged to manage their emotions effectively, which can lead to:
   a. Problems such as anxiety, depression, and irritability, which may cloud their decision-making clarity and negatively impact their relationships and overall well-being.
   b. Difficulties in communication and building strong team alignment, collaboration, and relationships, both professionally and personally.
2. Your leadership and staff may be overwhelmed with many other time-sensitive tasks that can lead to:
   a. Bad decisions, wrong choices, or a lack of insight and direction.
   b. A burnt-out workforce that can lead to poor performance, absenteeism, and retention problems.
3. Business priorities and goals may not be set appropriately, plans may be created with unrealistic expectations, or actions taken that have little contribution to the established goals.
4. Your leadership and staff may not have the capacity to lead others effectively. As a result, this can create distrust and an inability to promote a positive work environment.

Besides the issues identified above, there are many other potential impacts on the activities taken by process improvement teams or individuals when they are not being mindful as they engage various stakeholders to attempt to solve a burning problem or issue for their organization. Some of these may include:

1. Lack of leadership or wrong leadership support
2. Wrong project strategy selected
3. Wrong problem identified and resolved
4. No real connection between the process being analyzed and the problem experienced by the customer
5. Lack of attention to the non-verbal cues when gathering VoC
6. Lack of focus on people engagement
7. The scope of the problem is too broad to address adequately
8. Key critical input not considered, such as critical suppliers
9. Too little or too much focus on data
10. Lack of linkages between actions and outcomes

Hung Le: *A specific illustration of the above problems is an example I discussed in the Preface & Introduction, about my illness during a trip abroad. Granted, not all my problems could be attributed to a lack of mindfulness. Even when we have done our part, ensuring that all available and relevant facts are considered, there may be much that we don't know (and we don't know that we don't know it!). Consequently, bad decisions can be made. That's a fact of life. We must accept these occurrences mindfully and move on.*

When the head physician encouraged us to remain at his hospital for treatment of meningitis:

- *He assumed that I only cared about the number of similar cases his hospital had treated.*
- *He appeared unconcerned that his actions could impact me in a significant way – literally a matter of life or death.*
- *He was unaware of his recommendation's consequences for his own organization. If I died on his watch, the damage to his hospital's reputation would have been overwhelming, for I was a tourist, an American, and a volunteer for a charity mission! It would have been a public relations nightmare, to say the least.*
- *When we decided to follow our instincts and leave for another facility, my wife and I asked to rent an ambulance. What we got was an old ambulance truck that could have easily broken down during the seven-hour journey, actually putting both of us in further danger.*

Was this doctor putting me, a real (and quite ill) person, plus all my surrounding circumstances, at the center of his thinking? Was he considering my wife, my nationality, my purpose for being there? Or was he thinking in terms of bed space, medicine, and a treatment plan he'd been taught to follow in response to these symptoms, despite its ineffectiveness? And once I was delivered into better care, what about those patients who didn't have the information, finances, or ability to demand the same? Do they deserve any less consideration than I did?

There were quite a few other outcomes for the hospital's actions that they failed to consider. Any one of those outcomes could have been gravely adverse for me. But this situation was not unique. It was a lack of overall mindfulness, which permeates many organizations. Human compassion is key.

The present challenges we confront are overwhelming and complex. Our living and working environment is in a constant state of rapid change, measured in internet microseconds. Our organizations are continuously faced with new economic and resource limitations. We are surrounded by numerous technological devices that frequently generate excessive and irrelevant information, leading to information overload and a feeling of disconnection, which can overwhelm and isolate everyone. Those currently preparing for a career may discover that their intended path undergoes drastic changes by the time they are ready to embark on it. Lastly, the overwhelming multitude of voices and opinions on any given matter often leaves us uncertain about whom or what to trust or follow![1]

However, with challenges come new opportunities! Technological development has caused a rapid shift to a digital economy, leading to increased applications of artificial intelligence and machine learning, the Internet of Things, virtual and augmented reality, and healthcare technology, as well as the fast-growing implementation of renewable energy, to name a few.

Companies that can take advantage of these opportunities know how to create an environment that nurtures and promotes creativity, productivity, and compassion at all levels within the organization, not just at the leadership level but also as individual contributors.

In the new world economy, the human ability to be mindful (that is, to practice compassion, empathy, humor, intuitiveness, and gratitude) are factors that will continue to distinguish between a human and an AI entity. A human-centered organization will help leaders holistically address organizational problems by being inclusive and considering perspectives from all key stakeholders, including internal and external customers.

According to the Harvard Business School, this is what human-centered design brings: "A problem-solving approach that puts real people at the center of organizational excellence, enabling you to create products and services that resonate and tailor to your audience's needs."[2] Our personal belief is that only a human can develop emotional intelligence and not AI. Compassion will serve as the "glue" to elevate human performance through increased collaboration across business units and functions and leveraging AI to the fullest without fearing it! It will assist leadership and management in helping staff and employees connect with their higher purpose. Understanding of the concerns and needs of an organization's employees, customers, and stakeholders can promote trust and loyalty, leading to better engagement and commitment to the organization's goals.

Improved employee engagement can increase employee productivity, leading to better overall performance and lower turnover. Compassionate leaders will have the foresight to create an environment in which employees feel safe to express their ideas, take risks, and think creatively, leading to more innovative solutions and new growth opportunities. Finally, they help their organizations navigate difficult times and uncertainty, enabling them to respond more quickly and effectively to changing circumstances.

Both leadership and employees will develop an improved awareness of any given situation to allow them to dissect it better to understand the problem they should be tackling. Mindfulness gives us a better understanding of the vital few driving factors that help solve issues. Developing an unbiased mindset lets us identify the most appropriate solutions for the problem at hand.

In our view, this is what human-centered Lean Six Sigma should be – to bring the most value that can be delivered to your end customers while unlocking hidden potential from every single employee! That is the balancing act that should occur within any organization to thrive in a more competitive and complex business environment.

Time and Space and the Curious Mind

A long time ago, there was a Japanese Zen Master. People from all over sought his counsel and wisdom, which he was glad to share. One day, however, a man of some importance and wealth visited the Master. This man was accustomed to giving orders and being obeyed; he now commanded the Zen Master in the same tone. "I have come for you to teach me about Zen. Open my mind to enlightenment."

The Zen Master agreed with a smile and said they should discuss the matter over a cup of tea.

When the Master served tea, he began to pour a cup for his impatient guest. The tea rose to the rim as he poured it, then spilled over, leaking across the table and onto the man's expensive robes. Abruptly the man shouted, "You are spilling the tea all over! Don't you see the cup is full?"

The Master stopped pouring. He said to his guest: "You are like this teacup, so full that nothing more can be added. Come back when your cup is empty. Come back to me with an empty mind."

This often-told legend of the teacup illustrates that it takes humility to learn; we must set aside what we already know and open ourselves to new information and understanding.

As Corporate change agents, leaders, and consultants for many large corporations, we have observed some great leaders whose qualities go far beyond the scope of their roles and responsibilities. Besides delivering on their program and business commitments, these managers and leaders focused on bringing the best out of their people to achieve something bigger than themselves while nurturing an inclusive environment. Their cultures were human-centered naturally and by design! Yes, they indeed led by example. These compassionate individuals embraced their mission with passion and developed strong bonds with their colleagues across the enterprise and their community at large. They were constantly driven to excellence and always sought to make a difference in everything they touched.

We also observed leaders so overwhelmed that they had neither the time nor space to focus and listen deeply to themselves or others. Like the visitor in the Zen story above, their minds were already overflowing. How could they expect to generate connections with their stakeholders when they were so busy getting through their day-to-day activities? With so many distractions, how could they stay focused on the task at hand, or remain engaged in conversations with others without causing any misunderstandings? With increased responsibilities as the business grew more complex, could these distracted leaders trust their judgment and decisions and avoid making more mistakes? Were they able to deepen their understanding of their thoughts and emotions and maintain self-awareness?

As Victor Frankl once said: "Between a stimulus and a response, there is a space. In that space, it is our power to choose our response. In our response lies our growth and our freedom."[3] To create that space, one needs to be mindful of their thoughts and emotions.

Figure 11.2 illustrates the potential stimulus and response with and without mindfulness. Without mindfulness, one can merely react

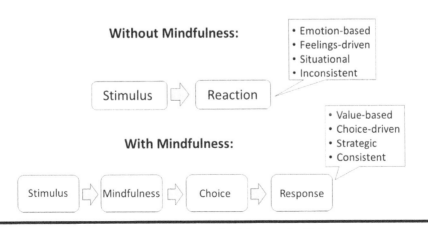

Figure 11.2 Stimulus and response with and without mindfulness.

automatically to current circumstances or situations, which can be emotion-based, feelings-driven, situational, and often inconsistent with one's values. One can bring clarity and positivity to many of one's life experiences with mindfulness. By being more aware of one's thoughts and feelings, one can choose a response to a stimulus that is value-based, strategic, and consistent with one's values. This will be a crucial leadership trait in today's fast-paced and dynamic business environment. Why? There exists no prescribed formula for success. Leaders make the most appropriate choices in any given situation for a given level of uncertainty.

How one goes about creating and sustaining a culture of excellence is a complex endeavor that will require sound decisions that must account for multiple dimensions. This requires a holistic and balanced view of all dimensions that must be considered in each situation. Leaders must be clear minded to make the most appropriate decisions while minimizing short-term and long-term risks. They must create space and time to make the most suitable choices and not just react to circumstances. That's the essence of mindful leadership.

Here are some tips for integrating mindfulness into your everyday leadership duties:

- Understand that your beliefs are sometimes driven by emotion. They are subconscious, automatic thoughts that can be illogical, invalid, or biased.
- Accept that your perception is limited. Your understanding of the situation is only one side of the story. Try to interpret the situation differently, change its meaning, or view it from another person's perspective.
- Your objective is to solve a problem rather than win the fight. Toward this end, you should try to understand new perspectives from others and look for ways to improve the effectiveness of your team's solutions to a given problem.
- Keep in mind that not everyone wants to reframe conflict as an opportunity – it's comfortable to ignore problems in the short term. However, leaving conflicts unresolved will lead you to similar situations in the future.
- Once you have shared your feelings about the situation, allow others to do the same.
- Have patience, and a sense of humor, about the fact that this paradigm shift will not happen overnight. Change can be challenging, and you may

take one step forward only to find yourself falling two steps back. Figure out where things went wrong and try again. It's okay to feel frustrated – just don't give up. Things will change for the better with time.

To Meditate or Not to Meditate?

Learning meditation is usually the first suggestion made for those who want to increase their mindfulness. However, while meditation can be a wonderful practice, it is not for everyone, or perhaps not for everyone all the time. That's all right; you need not "go into a trance" to be mindful. You can practice mindfulness in many other ways, simply by being fully engaged and aware of what you are doing.

Remember, meditation does not have to mean spending thirty minutes cross-legged, eyes closed, repeating a mantra. Anytime you center your thoughts, person and intentions, you are, in fact, doing a form of meditation. Just taking a deep breath to calm yourself is a type of meditating.

Practice doing everyday things with intention – walking, eating, talking to others. Make yourself pause throughout the day to appreciate what you have, doing a "gratitude check." We do so much multitasking, with so much noise coming at us, that it does take some practice, and purposefulness, to narrow our attention.

Hung Le: *I have a friend who sets an hourly alarm on her smartphone while she is at work. After she works for 55 minutes, the tone reminds her to engage in a five-minute mindfulness ritual. She might do one or more of the following things: take a short walk or stretch, think of something she's grateful for that day or something that made her laugh, practice deep breathing, or enjoy a small snack. She checks herself for focus. If something is distracting her from the task at hand, she imagines it going into a little box on a shelf in her mind. "I'll open that later," she tells herself. This easy practice keeps her mindful and makes work far more relaxed, enjoyable, and productive too! She gets a lot done during that other 55 minutes.*

As Jon Kabat-Zinn, an American professor emeritus of medicine and the creator of the "Mindfulness-Based Stress Reduction Clinic" explained: "Mindfulness is the awareness that emerges through paying attention on purpose, in the present moment, and non-judgmentally to the unfolding of experience moment by moment."[4]

Mindfulness emphasizes the importance of self-awareness, emotional intelligence, and the ability to remain present and focused on challenging situations. It is based on the principles of mindfulness, which involves paying attention to one's thoughts, emotions, and behaviors in a non-judgmental way. These mindful members of an organization can effectively navigate complex and fast-paced business environments by staying grounded in their values and principles, leading with compassion and empathy, and fostering a culture of growth and collaboration within the organization.

Microsoft CEO Satya Nadella offered this tip for success in a CNBC interview. Without mentioning mindfulness, he encompasses it perfectly. "There was never a time where I thought the job I was doing, all through my 30 years of Microsoft, that somehow I was doing that as a way to some other job...I felt the job I was doing there was the most important thing." He summarized, "Don't wait for your next job to do your best work."[5] Instead of being forward-minded, he focused on excelling in the role he had at the time.

At present, many US companies have implemented mindfulness programs for their employees, including Google, Aetna, General Mills, Proctor & Gamble, Target, AIG, Goldman Sachs, AstraZeneca, Fidelity Investments, and Salesforce, to name just a few. These companies offer a variety of mindfulness resources and practices, including guided meditation and stress management training, yoga classes, and on-site massages for relaxation.

The benefits of mindful leadership are numerous. Mindful leaders can better manage stress and maintain a sense of balance and well-being, allowing them to make more thoughtful and effective decisions. They also tend to be more effective communicators and better able to build strong, trusting relationships with their teams. Additionally, mindful leaders often foster a culture of creativity and innovation within their organizations, leading to increased productivity and success.

Becoming a mindful leader involves a lifelong process of learning and growth. It requires a commitment to continuous self-awareness, self-improvement, and the ability to remain grounded in one's values and principles in the face of challenges. Here are some critical steps to becoming a mindful leader:

1. **Develop mindfulness practices**: Mindfulness can be developed through various practices such as meditation, yoga, and journaling. Incorporating these practices into your daily routine can help you

become more self-aware and better manage your thoughts, emotions, and behaviors.

2. **Cultivate emotional intelligence**: Mindful leaders are often highly emotionally intelligent, which means they can understand and manage their own emotions and those of others. Developing your emotional intelligence can help you become more effective in your interactions with others and build strong, trusting relationships.

3. **Lead by example**: As a leader, your behavior sets the tone for your team. Leading by example can foster a culture of mindfulness and positivity within your organization.

4. **Foster a culture of growth and collaboration**: Mindful leaders often create an environment where team members feel supported and encouraged to grow and develop. You can create an environment where new ideas are shared and promoted by fostering a partnership and open communication culture.

5. **Stay grounded in your values and principles**: Mindful leaders are guided by their values and principles, which helps them to make difficult decisions with integrity and to navigate challenges with a sense of calm and perspective.

It's important to note that becoming a mindful leader is a continuous process. It requires a commitment to self-awareness, self-improvement, and being open to feedback and learning. The benefits of mindful leadership are significant; they go beyond the individual to the team, organization, and society. A mindful leader can create a positive and healthy work environment, fostering engagement, creativity, and productivity.

Here are some questions to consider about your role as a leader:

■ Are you aware of your emotions during stressful times?
■ Do you remain present and focused during challenging situations?
■ Are you able to make effective decisions, communicate well, and develop stronger relationships within your teams?
■ Are you able to "make room" for creative thinking and problem solving?
■ Are you committed to continuous self-awareness, self-improvement, and remaining grounded in your values and principles?

And here is a challenge for the truly brave-hearted. Pose these same questions about your leadership to the people or teams you lead (and let them respond anonymously, of course, so you'll get more honest answers). You might believe yourself an excellent communicator – but do others? Maybe you communicate extremely well to groups but not as well in one-on-one meetings. Or you might believe yourself to be open to creative thinking and problem solving, yet your team sees you as resistant to changes in technology. If you are truly committed to continuous self-awareness and self-improvement, mindfulness means recognizing and acknowledging when certain aspects of your leadership could use some attention. Being a leader doesn't mean you are pristine or flawless. We admire leaders who are accountable for their behaviors and strive to improve.

Managing the Self

Ask yourself the following questions:

■ What beliefs about yourself inhibit your best performance as a team member or leader?
■ What beliefs about others inhibit your best performance as you lead or work in a team?
■ How do you get in your own way? Do you sabotage yourself?
■ What damaging thoughts go through your mind as you work in a team?

The practice of understanding and managing oneself is an essential aspect of personal development and growth. It involves gaining insight into one's thoughts, emotions, and behaviors, and learning to regulate them to benefit oneself and others. This process can help individuals lead more fulfilling and satisfying lives and develop better relationships with others.

Self-understanding is the first step in the process of self-management. This can be achieved through various methods, such as introspection, journaling, and therapy. By gaining a deeper understanding of oneself, individuals can learn to identify patterns of behavior and thoughts that may hold them back and develop strategies to change them.

Once individuals understand themselves better, they can move on to the next step: self-regulation. This involves learning to manage one's thoughts, emotions, and behaviors to benefit oneself and others. Note, it does not mean suppressing or denying emotions, but instead, recognizing emotions simply for what they are, and stopping them from ruling one's actions. This can be accomplished through various techniques, such as mindfulness, cognitive-behavioral therapy, and meditation. Individuals who can self-regulate develop greater emotional intelligence (EI). They then respond to challenging situations more constructively and positively.

In addition to personal benefits, self-understanding and self-regulation also play essential roles in building solid relationships. When individuals better understand themselves and their motivations, they communicate more effectively and develop more meaningful and fulfilling relationships. They can also learn to empathize with others more effectively and respond compassionately to difficult situations.

We can only help others if we are conscious of the impact of our behavior. If we are conscious (mindful) of ourselves:

- We understand our beliefs, assumptions, perceptions, and actions and how they impact our interactions with others.
- We choose to interact with others in a deliberate, intentional manner rather than from an automatic response.
- We become more aware, conscious, and deliberate about our actions, intentions, and mindsets.

Understanding and managing the self is an ongoing process that requires consistent effort and practice. However, the benefits of this practice are numerous, including greater self-awareness, improved emotional regulation, and stronger relationships with others. Additionally, self-understanding and self-regulation skills are transferable to all aspects of our lives, helping us manage challenges in different areas, such as career, education, or finances. Investing time to understand and manage the self can help an individual lead a more fulfilling and satisfying life.

Questions for Discussion

1. What benefits could you derive from a lifelong practice of mindfulness?
2. How can you gain insight into your thoughts, emotions, and behaviors, and learn to regulate them to benefit you and others on your team?
3. What have you done in the past to improve your self-awareness, practice emotional regulation, and build stronger relationships with others?

Notes

1. Janice Marturano, Transforming Leaders into Mindful Leaders, Mindful Healthy Mind Healthy Life, 4/23/2019.
2. Landry, Lauren, *What is Human-Centered Design?* Insights, Harvard Business School Online, December 15, 2020.
3. Frankl, Viktor E., *Man's Search for Meaning* (1946), Beacon Press, Boston, MA, 2006.
4. Kabat-Zinn, Jon, *Mindfulness for Beginners*, Sounds True Publishers, Louisville, CO, 2016.
5. Nadella, Satya, CEO Microsoft, *Number 1 Tip for Career Success*, CNBC interview, accessed March 24, 2023. https://www.cnbc.com/2023/03/24/microsoft-ceo-satya-nadellas-no-1-tip-for-career-success.html.

Chapter 12

On Sustainability

> We are what we repeatedly do. Therefore, excellence is not an act, but a habit.
>
> **Aristotle**

Figure 12.1 shows many good behaviors that lead to increased health and fitness. You probably performed some of them already, but maybe not all. The ones you do perform are likely *habits*, something you have been doing regularly for so long that you hardly think about it. It's an ordinary part of your life and schedule.

Now think about the ones you don't perform as habits: perhaps exercise, meditation, mindfulness, or having your checkups. You know that these are valuable additions to your health. When you do remember to include them in your schedule, you're glad you did! But most of the time, you're ditching those options for other habits that are well-ingrained in you – like grabbing candy or a cigarette when you're stressed, binge-watching TV rather than taking a walk, or drinking soda instead of water. Such bad habits are usually hard to break because they are instantly gratifying, whereas "good" habits sometimes take a while to pay off in a noticeable way. If you want to truly make a change in your life, you must purposefully and intentionally (shall we say mindfully?) begin to include these behaviors in a routine until they, too, become habits, and you begin to reap the long-term rewards which provide the motivation to sustain your behavior.

There's a bit of a paradox here, though. Habits are our brain's own version of Lean Six Sigma. We form habits so that our brains and bodies don't have to expend energy deciding what to do every time a stimulus

DOI: 10.4324/9781003454892-12

Figure 12.1 We know what's good for us. Why don't we do it?

presents itself. That's why we can do things like walk and talk at the same time! We maintain habits that are rewarding, that get us something we want, that make life easier for us. This perfectly natural process only breaks down when our habits become outdated or maybe even destructive. Now, the problem is that we can look directly at a habit and recognize that it no longer serves its purpose yet continue it regardless. To our thinking, it's simply easier to continue doing what is familiar. Forming a new habit will take thought and energy, and possibly some sacrifice, and we simply aren't motivated to do it.

Sustaining behaviors in an organization is no different. The daily behaviors at every level of the organization are mostly habits. Some of these habits are good – otherwise there probably wouldn't be any organization at all. Some are sufficient but could be vastly improved. Some are quite wasteful – and we may even recognize them as such. But knowing a habit is wasteful doesn't change the fact that it is a habit, and it will continue until we replace it with something better. This doesn't mean simply "offering a better option," and hoping the organizational structure will choose it in the long term.

Instead, it means implementing the better option at every necessary level, letting every level know exactly why this new option is being presented and its purpose, then following up to ensure that the option is chosen until it becomes habit, while also ensuring that the new option really is consistently better by measuring results. Sustainability means the habit continues productively and is somehow rewarding to the performer – the new habit makes the task more enjoyable, profitable, satisfying, comfortable, or inspiring, just to name a few possible benefits.

Improvements must be sustainable and rewarding to become habits; otherwise, they are just irritants that interrupt work for a while and then are forgotten.

The Challenges to Organizational Cultural Change

We have observed many challenges to implementing an enterprise-wide change initiative. These are surprisingly common missteps:

1. **People assume an initiative must be very large, costly, and time-consuming.** That's far from true. Later, we will describe an efficient approach to changing an organization's culture.
2. **Initiatives are often limited to a tiny population in the organization (5 to 10%).** The implemented initiative could deliver real change in an organization's DNA except that most of the organization never knows it is happening – the initiative was only introduced and applied to one department, to one office, to one stratum of management.
3. **Organizations drift back to their old ways**. If there were changes that impacted an organization, these were small. If any significant impact resulted from such a change, many of the initiatives were not sustainable. Most organizations reverted to their old habits, meaning that those organizations must try to improve the same situation repeatedly. Sustainability requires effort: communication, follow-up, and

the certainty that everyone knows why something must be done this way (another reason why the entire organization should be included).

4. **Lean Six Sigma initiatives are implemented as an enterprise-wide initiative but not adopted by people in leadership or other pivotal positions.** Many tools and methods introduced as part of the initiative were meant for "others" in the organization rather than themselves. They felt that they couldn't use those tools and techniques.

5. **Lean Six Sigma was often run as a program or a function that did not apply to how everyday business was done**. This has caused different parts of our programs to be alienated from such initiatives. Often, these tools and methods were implemented in a narrowly specialized business area with overly rigorous improvement events, lean tools, and statistical methods that are often applied by black belts only. We do these intricate initiatives only when we must, and they are often very time-consuming.

Create a Culture Shift

To create a culture shift, one needs to understand that there is a difference between position and organizational influence. While a handful of leaders are at the top, most performers are at the grassroots level. One must recognize that top leadership can only *influence* an organization's culture. In contrast, culture is *defined* at the grassroots. To shift an organization's culture, what needs to happen is to transform people's behaviors at the grassroots level.

Whatever behaviors an organization would like to see in the culture, those behaviors are the change that needs to happen at the grassroots level. The effectiveness of any training conducted at the leadership level will depend significantly on how well and how much the leaders will apply concepts and practices from those trainings to impact or change the culture in an organization. So, any training that can change the behaviors of an organization will have to engage most of the population in an organization. As discussed earlier, the people being touched by an enterprise-wide change initiative often reside in small pockets within an organization. These pockets must be adequate to shift the culture of an organization.

Additionally, the kind of training intensively delivered often requires a considerable time commitment away from everyday work, which can distract from the day-to-day operations. We are proposing another training model that can be more effective in minimizing the time commitment away from everyday activities and maximizing training retention. We would call the traditional training model "high touch low frequency" versus the newly proposed training model "low touch high frequency." Getting more frequency provides an opportunity to reinforce new concepts, allowing individuals to internalize them, and eventually become a part of their daily work habits.

Appropriate behaviors are shaped by expectations that are derived from skill gaps. Then, proper training is provided to employees to close those skill gaps. But organizational behaviors are not "baked in" until employees can practice the concepts and tools learned from their training. It takes a long time for these behaviors to become sustained.

Shape Habits to Promote Excellence

Running a business is a team sport. You have got to have people in the right positions with the right skill set, and at the right time, so they can all work together to win games. Skilled athletes are indeed critical, but if they don't gel and work well together, then it could be chaotic. Imagine a team of top-notch athletes competing in a rowboat race. While all the athletes are top performers in their game, they will go round and round in a circle if they don't row together and sync up. So, you can have the best boat design, the best paddles, the best techniques, and the best athletes. If they don't align, sync up, and row together, they will lose 100% of their races.

The *right* people are an organization's best asset, not just any one individual filling a spot. This reminds us of a statement from Jim Collins's book, *From Good to Great*. He speaks of top-tier organizations that have disciplined people who have disciplined thoughts and can carry out disciplined actions. That is what it takes to win games. When you have the right people, you can have inferior processes and systems, and these people will look to improve them. In the long run, having the right people allows organizations to thrive in this fast-changing world. The flip side is that when you don't have the right people, you can have the best processes and systems in the world, and they will likely not use them effectively.

So, to start changing culture, consider the behaviors you must instill in your organization so that they can thrive and sustain over time. There must be clear expectations, adequate reinforcement, and strong motivation so that you can build those cultures, instill those behaviors in an organization, and then deliver on your goals. Employees must have the proper training, so they have adequate skills and tools to perform their job. But the most crucial part is that they must put what they learn into practice. Having the best tools and processes only means something if people use them effectively. It's not about best practice. What matters is that **you** practice.

A balance should always be maintained to achieve a specific objective. Anything left on its own at an extreme cannot work. For example, when we speak of having facts and data, we want our employees to gather relevant facts and data before they conclude or make any decision. But this does not mean that they spend excessive time collecting facts and data, which can lead to analysis paralysis and getting nowhere. However, at the opposite end of this spectrum, we must recognize that intuition is just as valuable, especially when it is developed through experience.

In the book *Blink* by Malcolm Gladwell, he described an exercise whereby a general was fighting simulated warfare against a large team of military analysts. Because the general was skillful at understanding situational awareness and had so much insight based on his experience, he could make swift decisions and win the simulated warfare even against such intelligent opponents.

Of course, relying on intuition alone can be a very dangerous proposition, as sometimes a slight variation could mean that what was done from experience before may not work this time. Solutions must be tailored and adapted to address the various nuances in any given situation. One should use facts and data to validate our insight based on intuition.

In another instance, when we say customer focus, it means that we should focus on the customer's needs first, but this does not imply that we should ignore and not focus on the processes critical to delivering on our customers' promises. We have observed that organizations focus on one aspect of the equation and ignore the other. We should focus on the customer and then ensure that we have good, efficient processes to deliver on those customers' promises.

Effectiveness requires doing the right things, not just focusing on efficiency. Ensure that you know what the right things are, then ensure that you can do those things the right way. So, efficiency for the sake of efficiency will only bring value if it is focused on the delivery of value for the customer. Sometimes there are considerable opportunities to work cross-functionally, and sometimes the organization should focus on optimizing its operation. Maximizing our operation is always easier because we have total control over it.

When there are handoff issues among the different functions, optimizing our operations alone cannot impact the overall flow of cross-functional processes. Sub-optimization will likely occur without looking at a situation in a holistic manner. Planning without action is just a plan. And action without a plan is a plan for failure. So, in effect, you must plan **and** act on that plan.

Foundation of Continuous Improvement (CI)

Excellence can be defined as an attribute or a measure against some expectations derived from desired outcomes. Determining what both internal and external customers value, and how to deliver them most efficiently, is one of the main goals of an organization. The foundation of continuous improvement is based on the DMAIC model (see Figure 12.2; please refer to Chapter 5 for a more detailed description of the model), which stands for Define, Measure, Analyze, Improve, and Control. Many different models exist but are set up to achieve similar objectives. Overall, these DMAIC phases form a scientific approach to problem solving, where one develops the hypothesis, gathers appropriate data, and uses it to prove or disprove your hypothesis so that, in the end, the best solution can be identified and implemented to resolve the current problem.

One of the problems we have witnessed with many Lean Six Sigma implementations was their concentration within small groups that account for between 5% and 10% of an organization's total population. They are made

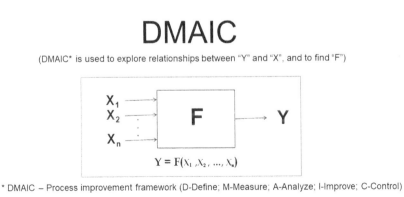

DMAIC

(DMAIC* is used to explore relationships between "Y" and "X", and to find "F")

$$Y = F(X_1, X_2, ..., X_n)$$

* DMAIC – Process improvement framework (D-Define; M-Measure; A-Analyze; I-Improve; C-Control)

Figure 12.2   Foundation of continuous improvement – Y = F(X).

up of black belts and green belts and continuous improvement practitioners who all speak their lingo. The complaint was that these improvements took a long time to realize any gains. But many of the gains were fairly insignificant. Occasionally, there might have been some significant gains, but they could have been more sustainable. Also, people who participated in some of these improvements after completing their certifications are nowhere to be found nor participate in any future improvements. It was as though they had completed their degrees, and there was nothing more to be done. So, the knowledge they have gained from one iteration by no means allows them to become an expert or contribute value in terms of solving organizational problems in the future.

When the CI practitioners participate in an improvement, they are often required to participate in a complete DMAIC lifecycle project. But this is hardly what CI is all about. Many of the activities conducted as part of the DMAIC lifecycle can be applied and do not require these CI practitioners to complete an entire DMAIC lifecycle.

For example, the Define Phase is about identifying issues or clarifying opportunities for improvement before problem solving, or it is designed to ensure that you have diagnosed the right problem before taking it on. This practice can be applied anywhere – in a meeting or at a workshop where one might hear about a situation someone wants to tackle. The question is, has this team thoroughly understood issues surrounding the problem, and what opportunities lie for improvement before investing scarce resources to take on the problem in the first place?

The Measure Phase is about targeting the relevant facts and data to understand the current situation or to focus on monitoring the heartbeat measures of the process. How valuable would it be for a team to speak up or challenge whether one should take on a problem and question whether the correct facts and data were gathered to validate the current situation before investing time and resources in it?

The Analysis Phase is about recognizing the causes of the problems and the potential impact of new solutions or innovation. It also addresses the relationship between process inputs and outputs. Without an in-depth understanding of the root causes, the team will be chasing after ghosts and likely only managing the problems' symptoms. Consequently, if the team has not addressed the issues or problems being observed, the same problem will recur.

The Improve Phase is about developing effective and innovative ways to get impactful results or pilot solutions with a bias for action. How often do we implement a full-scale solution even before we pilot it to validate that the solution has proven to work? Or how often do we jump into the first solution in mind and go ahead and implement and the resulting outcome is that the implemented solutions are less effective than one thought?

Finally, the Control Phase is about sustaining solutions and innovations and using leverage to maximize long-term benefits or to continuously strive to learn, improve, and iterate for continuous improvement. What we have observed over many years was that sustaining implemented solutions was a big issue.

As discussed earlier, many other improvements have resulted in minimal gains. If they result in a huge impact, they often cannot sustain over time because of various reasons such as reorganization, turnover, or a change in the business and market dynamics. By unbundling all these DMAIC activities, there is a lot to be gained by getting these practices to be part of everyone's behaviors which can then contribute to the work or activity they are engaging in (see Figure 12.3). As Aristotle rightly stated: "Excellence is not an act but a habit!"

To address some of the shortcomings of Lean Six Sigma initiatives, one of the first vital steps to prepare for such an implementation is to create a foundation that enables culture change. This will occur when the frontline

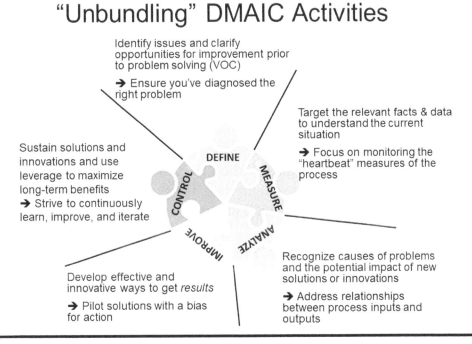

Figure 12.3   Unbundling of DMAIC activities.

staff is aligned and there is a strong commitment to teamwork. Alignment can be achieved by setting clear goals and expectations, developing employees, and transforming them to behave in specific ways that will allow maximum collaboration. As part of the building block to create a culture of excellence, it's worth highlighting the three "R's" in Relate, Repeat, and Reframe framework.

This has been discussed at length in the book by Alan Deutschman, *Change or Die*.[1] In this book, Deutschman discusses patients who went through successful bypass surgeries. At the end of the very successful bypass surgery, the head physicians said that if you don't change your lifestyle, you will die!

Sure enough, within 12 to 16 months, about 80% of these patients revert to their old lifestyle, knowing they will die! To impact these statistics, a team was formed to include a care nurse, a physician, and a case manager who organized weekly meetings with these patients to discuss what they were going through and continue to reinforce some of the things that would help them see how their actions negatively impacted others. To make the long story short, this community of practice could engage the patients and relate to them in a way that empathizes with their situation. Specific behaviors were repeated and reinforced, making them more rewarding. And eventually, they were able to get these patients to see things in a different light. As a result, reframing how those patients viewed their health. Once they saw things differently, their behaviors changed. Following this study, within 12 to 18 months (about one and a half years), over 80% of the patients could live a new lifestyle and act and behave differently. They were able to sustain their new lifestyle going forward.

The traditional training mode for Lean Six Sigma tends to be intensive. It requires a significant time commitment away from work (anywhere from three days to two weeks), making it very challenging. The retention rate for this kind of training is not optimal, either; it's rather like cramming for a test. In the end, one can pass the test right then, but there is minimal retention of all the content delivered during the training. A month later, how much has really been learned?

That training model is what we would call "high touch and low frequency": high touch because the face-time commitment is very high, and low frequency because it only occurs once or twice a year. Refresher courses and follow-up would greatly improve retention.

Hung Le: *My friend and her husband were excited to take a ballroom dancing course. But when they began the class, they found their instructor's teaching style to be quite unhelpful. The instructor and his partner would show the couples the entire dance sequence for the mamba or the tango – some of which could be ten to fifteen steps altogether – and then say, "Now, you do it." The couples stumbled through ineffectively. The instructor and partner would demonstrate again, the entire sequence of moves, and then say, "You, see? That's how you do the dance." Some of the couples were able to mimic most of the moves, or at least the first few, and the instructor considered this to be sufficient. That was the end of that dance lesson; the next week they moved on to another dance, with no better methodology in the instruction. Within a matter of days, my friend and her husband could not remember anything about the dances except the starting position. Naturally, I had questions. "Why didn't you ask him to break it down for you? Why didn't you request that he teach the dance one step at a time?"*

"We did," my friend replied to me. "But he didn't know how to break them down. I think he's been dancing for so long; he thinks that the whole sequence is just one step. And he couldn't figure out why we didn't see it that way, either." The new training model aims to avoid information overload and focus on sustaining incremental progress. Such a model is something we would call "low touch and high frequency." Low touch is important because it focuses on a few key concepts one can learn and practice. And high frequency allows learned concepts to be reinforced repeatedly so that retention can be improved.

Lastly, we unbundled core Lean Six Sigma concepts and related it to what people see and do daily. This will help with understanding and make it practical to apply in people's everyday activities. Shifting the culture of an organization helps to keep in mind the sets of critical behaviors that you want people to possess so that they can apply and put them into practice, thereby transforming them into a set of habits.

The Building Blocks

Building a culture of organizational excellence will require a systematic approach to ensure that an organization achieves its performance goals and that those achievements are sustainable. Organizations often need to understand the maturity of their processes and organizational readiness to implement *before* implementing such methodologies as Lean and Six Sigma.

One may not be able to improve those processes when there are misalignments within an organization, when many of its processes need to be better defined, or data used to determine the state of those processes must be better understood. We will soon discover that people revert to how they did things in the past because they lack a firm, organization-wide foundation. As in the example of the champion rowers, everyone needs to row at the same time, in the same direction.

Alignment goes hand in hand with having the right people on the team. As discussed in Jim Collins's book *From Good to Great,* it is essential to have disciplined people with disciplined thoughts who can carry out disciplined actions. More than having good processes is necessary. Without having the right people, it is likely that many of those processes considered to be best practices may not be used for the benefit of the organization.

Therefore, it is vitally important to have the right people with goals established and communicated so that there is alignment from the top down throughout the organization. These first two steps are necessary for improving the organization's performance without tools like Lean and Six Sigma. Remember from our discussion of process improvement, you cannot improve an unstable system. The first order if you have an unstable system is stabilization. We do this through aligning teams within an organization to the point where you have a stable working system.

The next step is to continually assess a set of goals and expectations to determine where adjustment may be needed. Participation is also a crucial step where everyone in the organization should consider how well they are aligned to support their mission. The next few steps are about getting a more detailed look at where one can improve by linking actions to outcomes. And then introduce how processes can be enhanced with a more in-depth understanding of data and variations, improve operations, and sustain them over time. The last step is critical because, given that there has been so much effort put in place to establish organizational excellence, you will want to ensure that it can be sustained over time. Mindful leadership is crucial to ensure that everything gels together, and you have an optimal system to work with.

Overall, it helps to think and lay out a road map to achieve organizational excellence and build a culture around it. We have tested an approach that has proven to work in a small department and were able to scale it up to a much larger organization. Figure 12.4 summarizes the road map. Please refer to Chapter 4 for a more detailed description of the road map.

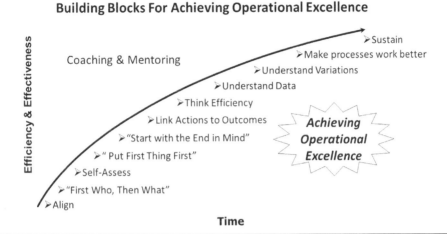

Building Blocks For Achieving Operational Excellence

Figure 12.4   Building blocks for a culture of excellence.

Leadership Engagement

From an enforcement perspective, it helps the entire team, including management, to be involved in this training. The reason is that everyone on the team hears the same message and has the same understanding of the approach and the tools and concepts taught. Concepts learned at each session are then put to practice at every opportunity to reinforce them.

This is how management and leadership can help build a culture of excellence. Although this may seem obvious and easy to implement, when an organization faces numerous operational challenges in a fast-changing business dynamic, people-development aspects tend to fall by the wayside and the focus moves to executing the task at hand.

Considering this problem, it's easy enough to see why this happens. People are probably the most adaptable part of a process. When a challenge suddenly arises, the fastest and simplest thing to do is tell the people to adapt, to make it work, to find a way around the difficulty. Often, a behavioral change "solves" the problem almost immediately, at least on the surface. We like to see things fixed fast (instant gratification!) and this rewards us, so that we may ignore the toll it takes on the people who must cope with the compounding waste, stress, ineffectiveness, and unpredictability of the process.

There was an administrative assistant for the manager of a department that was just barely meeting quotas. The assistant stated that his job was to "spend all day apologizing for my boss's behavior and fixing his mistakes." The assistant was burned out and looking for another job – he was but one in a long line of similar workers who had been hired to cope with a poorly-trained, unmotivated, and distracted manager. The fact that the department kept meeting its requirements at all was thanks to the "cleanup" efforts of these assistants. Making a list of all the types of waste in this scenario…well, it could take a while. While the skills of these assistants were wasted in clean-up efforts, the manager was clearly in the wrong job or had the wrong training for what he was meant to do – and perhaps the process itself needed improvements that could have reduced stress and misunderstandings. Small wonder that the department constantly struggled to meet its minimal requirements.

Although skipping over people-development aspects may seem like the right thing to do in the short run, without an intentional effort to develop the essential habits to help drive organizational excellence, the organization will not be able to achieve its long-term goal of creating and sustaining a culture of excellence. As discussed earlier, culture is defined at the front line, and leadership can only help shape and influence this culture. This is a prime opportunity for leadership to engage the workforce and help shape it in a way that could benefit the organization in the long term.

Observed Changes in Behaviors and Habits

Common sense is not common practice.

Stephen Covey

As discussed, problem solving using Lean Six Sigma accounts for a small percentage of an organization's time. This also implies that the population participating in process improvement is limited to small pockets within the organization. Therefore, achieving a shift in behaviors will be the key to unlocking the potential of everyone in the organization. When people develop a deeper appreciation of customer needs, it will go a long way toward assisting the organization in succeeding in this fast-paced business environment. When people recognize that excess inventory is a waste, when they recognize the need to be more productive in a meeting, or they realize that asking the why question multiple times to uncover the root causes is essential, when they realize that someone who is at the receiving end of the products or services they work on is their primary customer, or that they are being challenged to be more engaged and share ideas, or they have a heightened awareness of the cost of poor quality, and many other such behavior changes, a culture shift has occurred!

One of the critical successes of this approach is that employees who went through the low-touch high-frequency training could put their learning into practice instantaneously. They were able to relate to those concepts in the classroom. They were able to translate them into how they would do things differently in their everyday activities. They were pretty inspiring! We are not suggesting that this approach should solely focus on shifting mindsets and behaviors, but this should supplement how the process improvement teams are engaged. There will always be opportunities to execute process improvement using the entire DMAIC lifecycle. Still, the point is that these are few and far between, and they can hardly change the culture by just focusing on a few strategic projects.

We observed that after people went through this training, they had a deeper appreciation for more advanced tools and were eventually engaged in more in-depth training to either become a green belt or black belt in Lean Six Sigma of their own choice! It was indeed a "pulled" versus a "pushed" model! Note the actual transformation of an organization toward a culture of excellence where everyone is looking to improve their game continuously. When you see everyone in your organization questioning the status quo, looking for different ways to make things faster, cheaper, and better, working together, and aligning to the organizational goals to drive excellence within the organization, you see true transformation occur.

Critical Success Factors

Below are the tips and lessons learned that will serve as critical success factors to drive and sustain excellence within an organization:

- Reinforce key LSS concepts learned:
  - Coach and mentor your staff.
  - Focus on not just the WHYs and the WHATs but also the HOWs.
  - Provide regular feedback.
- Drive CI (Continuous Improvement) deeper into the organization:
  - Improve LSS training effectiveness and expand training offerings by engaging every part of the organization.
  - Expand the reach of CI strategic improvement projects.
  - Encourage students to apply the concepts to an everyday problem-solving approach.

- Promote CI and engage all levels of the organization:
  - Expand participation of CI practitioners at a Cross-Functional Monthly CI Forum/Community of Practice (CoP).
  - Engage Managers and Directors to reinforce a CI mindset – Put a stronger emphasis on the Whys and the Whats as well as the Hows.
  - Empower the frontline employees to continue to make incremental changes and pilot new solutions and innovations themselves.

Questions for Discussion

1. What is the penetration of Continuous Improvement engagement within your organization?
2. Is your leadership and management engaged in Fundamental Continuous Improvement Training? Is there enough support from your leadership?
3. Is there a CoP established? Are all levels of the organization engaged?

Note

1. Deutschman, Alan, *Change or Die: The Three Keys to Change at Work and in Life*, Harper-Collins, New York, 2009.

Conclusion

As we defined in Chapter 1, a human-centered organization is "an organization that seeks to deliver the most value to the end customer while unlocking hidden potential and leveraging talents from every single employee to thrive in this very dynamic, fast-paced, and complex business environment." To achieve this, we aim to help prepare every single employee to become an effective problem solver, supporting their department's goals and aligning with the organization's broader goals. We want employees to be more effective, efficient, and inspired by having access to the right tools and techniques, improving their chances of success. This should serve as a solid foundation for continuous improvement, thereby enabling excellence to thrive within the organization.

We offer the following reminders as you embark on your journey. To create and sustain a culture of excellence for your organization:

- Align your priorities with the broader goals of your organization. It is necessary to prioritize challenges and execute them so that short-term gains can be achieved, and sustained, to support longer-term goals.
- Put people first. People perform the work! Do not forget that. Implement tools and processes to help people achieve better outcomes. Spend more time leading and inspiring people, not just managing the work. Once people are inspired, they will "own" the work, manage it and produce outcomes that will likely exceed customer expectations.
- Use the Building Blocks for Operational Excellence (Figure 4.2) roadmap as a guide. It will help you assess where your organization is from a maturity perspective so that appropriate steps can be taken to help improve over time. Be sure to leverage past initiatives and build upon them.

DOI: 10.4324/9781003454892-13

- Move from just teaching tools to instilling appropriate behaviors that can change organizational behaviors at the grassroots level. Intentional change takes more than just willpower. You will need support from leadership and colleagues to drive meaningful change. Forming a Community of Practice is vital to institutionalizing change.
- Learn the tools well. Be insightful in solving problems. Learn how and where to tap, just like the consultant in the story told in Chapter 4! Your valuable contribution is to know where to tap to be an effective problem solver – You don't get paid for just tapping! Do not just follow the crowd; do not use the tools and techniques as a "cookbook." They are there, so you can apply them where appropriate. Apply them at the right time and place to get the most value out of tools.
- Start mindfulness practices and encourage employees to do so. Embed mindful practices into every employee's work activity to help cultivate a culture of growth and collaboration.
- Continue to put your learning into practice and "sharpen your saw"! Make time and effort to continue life-long learning and self-improvement.

We hope that you have benefitted from reading this book. Hopefully, we have inspired you to take concrete actions and apply what you have learned to improve your organization or personal life. We understand that culture change can be incredibly challenging, but it is ultimately achievable and can lead to significant benefits for everyone. And it starts with you!

We wish you all the best!

Hung Le and Grace Duffy

Index